Artificial Experts

Inside Technology

Wiebe E. Bijker, W. Bernard Carlson, and Trevor Pinch, editors

Artificial Experts: Social Knowledge and Intelligent Machines
H. M. Collins, 1990

Viewing the Earth: The Social Construction of the Landsat Satellite System
Pamela E. Mack, 1990

Inventing Accuracy: A Historical Sociology of Nuclear Missile Guidance
Donald MacKenzie, 1990

Artificial Experts
Social Knowledge and Intelligent Machines

H. M. Collins

The MIT Press
Cambridge, Massachusetts
London, England

This book was set in New Baskerville by The MIT Press and printed and bound in the United States of America.

Library of Congress Cataloging-in-Publication Data

Collins, H. M. (Harry M.), 1943–
Artificial experts: social knowledge and intelligent machines/
H.M. Collins.
p. cm.—(inside technology)
Includes bibliographical references.
ISBN 0-262-03168-X
1. Artificial intelligence—Social aspects. 2. Knowledge, Sociology of.
3. Expert systems (Computer science) I. Title. II. Series.
Q335.C54 1990
006.3—dc20 90-5541
CIP

Contents

Preface and Acknowledgments

What is knowledge? How is knowledge made and moved around? Over the last twenty years these questions have been explored within social studies of science. When one tries to put knowledge into a computer, the questions present themselves in an acute and well-defined form. That is why artificial intelligence research is a natural living laboratory for *the science of knowledge.*

When I set out to write this book in 1987, I had a clear idea of what I wanted to say. That idea, with the usual modifications, has turned into parts II, III, and IV of the book. Part I has taken me by surprise. It began to grow when I realized with a shock that the argument about the limits of asocial machines was so powerful that there ought to be no such thing as a pocket calculator. Part I shows how pocket calculators work.

The second part of the book includes ideas developed in five papers that have already been published. I thank the editors and publishers for allowing me to reuse some of the material from the originals. The first of the papers to be published was also my first attempt to contribute to the literature on expert systems. This paper is, "Where's the Expertise: Expert Systems as a Medium of Knowledge Transfer," in M. J. Merry (ed.), *Expert Systems 85,* Cambridge University Press (1985), 323–334. It was the favorable and generous treatment that this paper received at the British Computer Society Specialist Group in Expert Systems Annual Conference in 1985 that encouraged me to press on.

The other four published papers are: "Expert Systems and the Science of Knowledge," in W. Bijker, T. Hughes, and T. Pinch (eds.), *New Directions in the Social Study of Technology,* MIT Press (1987), 329–348; "Expert Systems, Artificial Intelligence, and the Behavioral Co-Ordinates of Skill," in B. Bloomfield (ed.), *The Question of Artificial Intelligence: Philosophical and Sociological Perspec-*

tives, Croom-Helm (1987), 258–282; "Domains in Which Expert Systems Might Succeed," *Third International Expert Systems Conference,* Oxford, Learned Information (1987), 201–206; and "Computers and the Sociology of Scientific Knowledge," *Social Studies of Science*, 19 (1989).

The ideas in the book also have been tried out in embryonic form at a number of conferences and seminars. In almost every case I have gained immensely from the thoughtful and critical remarks of the audience. I would like to thank members of the audiences at: The Workshop on New Directions in the History and Sociology of Technology, Technische Hogeschoole Twente, 1984; The Annual Meeting of the British Computer Society Specialist Group on Expert Systems, University of Warwick, 1985; The History and Philosophy of Science Seminar, The Hebrew University, Jerusalem, March 1986; The Alvey/SERC conference on "Explanation in Expert Systems," University of Surrey, March 20–21, 1986; ICL Association of Mainframe System Users—Scientific Users Group, Strand Palace Hotel, London, April 17, 1986; The Conference on Technology and Social Change, Centre of Canadian Studies, University of Edinburgh, June 12–13, 1986; The XI World Congress of Sociology, New Delhi, India, August 18–22, 1986; The IVth Annual Meeting of European Association for the Study of Science and Technology, September 29–October 1, 1986, Strasbourg; The Joint Meeting of the History of Science Society, Philosophy of Science Association, Society for the History of Technology, and Society for Social Studies of Science, Pittsburgh, October 23–26, 1986; The University of Aston, Technology Policy Unit, Seminar, November 27, 1986; The Hewlett-Packard, Bristol Research Centre, Seminar, January 30, 1987; The Conference on Cognition and Social Worlds, Department of Psychology, University of Keele, April 6–8, 1987; The University of York Open Lecture series, May 12, 1987; The Third International Expert Systems Conference, London, June 2–4, 1987; The Tuesday Meeting at Xerox PARC, Palo Alto, California, October 20, 1987; The Seminar of the Department of General Sciences, University of Limburg, Maastricht, The Netherlands, April 18, 1988; The Conference on New Technologies, Institute for Advanced Studies in the Humanities, University of Edinburgh, August 18–21, 1988; The Seminar of The Department of History and Philosophy of Science, University of Cambridge, October 27, 1988; The Xerox PARC, Palo Alto, Seminar on November 15, 1988; The 87th Annual Meeting of the American Anthropological Association, Phoenix, Arizona, November 16–19, 1988; and The Workshop on "The Place of Knowledge" at The Van Leer Jerusalem Institute, May 15–18, 1989. There are too many people who made useful comments either from the floor or in less formal discussions afterwards for me to mention them all by name.

Outside of these occasions many colleagues have given me ideas and suggestions; a few will find their names in the text. Those who read all or parts of the first draft of the manuscript and gave me useful advice include Jay Gershuny, Joanne Hartland, Rudolf Klein, Helene Laverdiere, Gerard de Vries, and Sherry Turkle. Wiebe Bijker, David Edge, Jean Lave, Jean-Marc Levy-Leblond, Barry Lipscombe, Don McCloskey, Trevor Pinch, Steve Shapin, and Lucy Suchman have given me stylistic or technical help and/or encouragement beyond what anyone has a right to expect, while Caroline Arthurs offered invaluable advice on points of English style. All remaining mistakes and infelicities are my responsibility.

I am grateful to my collaborators Rodney Green and Bob Draper, who worked with me on building the small expert system described in Part III of the book. Bob Draper showed immense patience and tolerance in the face of unreasonable demands on his good nature. This book could well be dedicated to *skillful* people everywhere. The laboratory technicians who actually make our science work are among the least appreciated contributors to our cultural world.

The Science Policy Support Group provided a small grant that enabled me to build the expert system, but no other grant moneys supported this research. On the other hand, Lucy Suchman and Xerox PARC provided me with an office and equipment to begin work on the book in a congenial atmosphere in an environment conducive to thinking about computers. Both Xerox and the University of Bath helped with the financial arrangements for this trip. The University of Limburg was my host during the final revision of the manuscript, and I thank Gerard de Vries for making possible my trips to the charming town of Maastricht and for providing, with Wiebe Bijker and Guy Widdershoven, a stimulating discussion group.

I would like to thank my students at Bath for listening to my ideas with tolerance at a stage when they were not really ready to present to a class. It was during one such class that the "calculator problem" struck me. Finally, I am grateful to my children, Joe and Lily, for being an incomparable source of examples of action both machine-like and nonmachine-like.

I

What Computers Can Do

1

Artificial Brains or Artificial Experts?

My subject is *knowledge science.* It is the study of what communities know and the ways in which they know it. Individual human beings participate in knowledge communities but they are not the location of knowledge. Rather, the way that individuals reflect the knowledge of communities is a topic for analysis within knowledge science.[1] Knowledge science looks at how knowledge is made, maintained, disputed, transformed, and transferred.[2] Artificial intelligence is a natural field site for knowledge science because intelligent computers appear to channel and constrain what is known by knowledge communities into well-defined, discrete, asocial locations. Though early claims were overambitious, there are intelligent machines and they are getting better. Yet the existence of any intelligent machine seems to contradict a basic premise of knowledge science because a machine is not a community or a member of society. What better starting point could there be?

The early misplaced confidence of the proponents of artificial intelligence is easy to understand. It is precisely analogous to the misplaced confidence of rationalist philosophers of science and, I suspect, has fed upon it.[3] If science was, at heart, a logical, individualistic method of exploring the world, then the computer, a quintessentially logical individual, could start with arithmetic, graduate to science, and eventually encompass much of human activity. In the last two decades, however, science has started to look rather different. Detailed empirical studies of the way scientists make knowledge have given us a picture of science that is equally far from philosophical and common-sense models. Building scientific knowledge is a messy business; it is much more like the creation of artistic or political consensus than we once believed. The making of science is a skillful activity; science is an art, a craft, and above all, a social practice.

Starting from this viewpoint, the prospect seems distant of making intelligent, problem-solving machines. If science, the paradigm case of human problem solving, turns out to be messy, crafty, artful, and essentially social, then why should tidy logical and isolated machines be capable of mimicking the work of scientists? Still less should they be capable of doing the more obviously messy work of the rest of us.

The history and sociology of scientific knowledge has shown that scientific activity is social in a number of ways. First, when a radically new experimental skill is transferred from one scientist to another it is necessary that social intercourse take place. No amount of writing or talking on the telephone appears to substitute for visiting and socially rubbing up against the person from whom you want to learn. We can contrast two models of learning: an "algorithmic model," in which knowledge is clearly statable and transferable in something like the form of a recipe, and an "enculturational model," where the process has more to do with unconscious social contagion. The algorithmic model alone cannot account for the way that scientific or other skills are learned. My own study of the transfer of knowledge among laser scientists illustrates this point; I found that scientists who tried to build a radically new type of laser—a TEA-laser—while working only from published sources were uniformly unsuccessful (Collins 1974, 1985).

A second way in which science is social is that conclusions to scientific debates, which tell us what may be seen and what may not be seen when we next look at the world, are matters of social consensus. Whereas the formal model of seeing—the pattern recognition model as we might call it—involves recognizing what an object really is by detecting its distinguishing characteristics, the enculturational model of seeing stresses that the same appearance may be seen as many things. For example, there is a well-known photograph that can be seen as the face of Christ or Che Guevara, whereas it is claimed to be a picture of a snow-covered mountain range in China. Sometimes viewers see it as quite other things including an abstract black-and-white pattern of splotches. The question "What is it really?" cannot be answered. No amount of ingenious pattern recognition programming would reveal the truth. There is no algorithm for recognizing the pattern.

Nevertheless, in saying that it can be seen as an image of Christ or Che Guevara, I am saying something true about how our culture sees the image. For example, in the West it is easy to persuade people in a classroom that it really is a face. Students who cannot see the face come to believe that it is their fault. They feel inadequate for not being able to see what their colleagues can see so clearly. Using routine classroom techniques (Atkinson and Delamont 1977), a group of students can be made to act like a small scientific community. The nearest analogy is the historical and continuing debate about what is real and what is artifact when you look through a microscope. Like the face, scientific facts do not speak for themselves. Disputes in science are not settled by more and more careful observation of the facts; they are settled by broad agreement about what *ought* to be seen when one looks in a certain way at a certain time and location. Thereafter, anyone who looks and does not see what everyone agrees ought to be seen is blamed for defective vision (or defective experimental technique). This process is illustrated in earlier work on the controversy over the detection of gravitational radiation (Collins 1985 and a host of other detailed field studies).[4]

A third way in which science is social is in what one might call the routine servicing of beliefs. An isolated individual, having no source of reference against which to check the validity and propriety of perception may drift away from the habits of thinking and seeing that make up the scientific culture. Again, the social group is the living reminder of what it is to think and act properly, correcting or coercing the maverick back onto the right tracks. Thus, learning scientific knowledge, changing scientific knowledge, establishing scientific knowledge, and maintaining scientific knowledge are all irremediably shot through with the social. They simply *are* social activities.

What is true of scientific knowledge has long been known to be true for every other kind of human cultural activity and category of knowledge. That which we cannot articulate, we *know* through the way we act. Knowing things and doing things are not separable. I know how to speak through speaking with others, and I can show how to speak only through speaking to others. Changing the rules of speaking is a matter of *social* change; it is a matter of changing common practice. If the rules of speaking change, then I follow along with the others, not because people tell me what to do but because in living with others—in sharing their "form of life" (Wittgenstein 1953)—I change with them. I will change what I know about how to speak, not as a matter of choice, not as a matter of following a consciously appreciated rule, not at the level of consciousness at all, but because in doing what others do I will find that I know what they know. In knowing what they know, I will do what they do. This is true of speaking, writing, plumbing, plastering, practicing medicine, and discovering subatomic particles. To put the issue in its starkest form, the locus of knowledge appears to be not the individual but the social group; what we are as individuals is but a symptom of the groups in which the irreducible quantum of knowledge is located. Contrary to the usual reductionist model of the social sciences, it is the individual who is made of social groups.

Now think of a computer being tested for its human-like qualities. Let it be subjected to a "Turing Test" (see especially chapters 13 and 14) in which it must engage in written interchanges so as to mimic a human. The computer is at the wrong end of the reductionist telescope. It is made not out of social groups but little bits of information. What will it not be able to do by virtue of its isolated upbringing? Consider a more familiar example.

A foreign agent, of the kind one sees in the movies, has to pass a kind of Turing Test. Imagine a spy, a native of London, who is to pretend to be a native of, say, Semipalatinsk. The agent has learned the history and geography of Semipalatinsk from books, atlases, town guides, photographs, and long conversations with a defector who was himself once a native of the town. He has undergone long sessions of mock interrogation by this defector until he is word perfect in his responses to every question. His documents are in order and he has a story that explains his long absence from the town. In the films, the British agent goes to the USSR and begins to spy. He is picked up by the KGB and interrogated; the value of all those hours of training are revealed as he answers his captors' questions. As in the Turing Test, the problem for the interrogators is to distinguish between real accomplishments and an imitation—between the spy and a real native of Semipalatinsk. The moment of crisis occurs for our hero when an interrogator enters who is himself a native of the town. At this point nothing will save the spy except a distraction, usually extraneous, which brings the interrogation to an end. However good his training, we know that the spy will not survive cross-examination by a native of Semipalatinsk.

The reason we know he will not survive that final cross-examination is that, however long the spy's training, he cannot have learned as much about Semipalatinsk as a native would have learned by living there. There is a very great deal that can be said about

Semipalatinsk, and only some of it can have been said during the training sessions. The spy will be able to make some inferences beyond what he has been told directly (for example, he will be able to form some brand-new sentences in the language), but he will not have learned enough to make all the inferences that could be made by his native trainer or his native interrogator. A native learns about Semipalatinsk by being socialized into Semipalatinsk-ness and there is much more to this than can be explicitly described even in a lifetime.[5] Thus the trainer, competent native that he was, cannot have completely transferred his socialization to the spy merely by talking to him for a fixed period. Photographs and films will help, but all of these are merely different abstracted cross sections of the full Semipalatinsk experience. Willy nilly, the trainer must have talked about only a subset of the things he could potentially describe, and even if the spy has absorbed all his instructions perfectly, he cannot know everything that the trainer knows, nor everything that the native interrogator knows.

The native interrogator will ask questions based on his own socialization—again, he can only ask a small set of the potential questions—but there is a good chance that during the course of a long interrogation he will ask a question the answer to which covers details that the spy has not encountered, or turns on an ability to recognize patterns that he has not seen, or requires an inference that he is not in a position to make. Is there an area of the town—near the river, perhaps, or going toward the forest, or just beyond the tanning factory—that is quite distinctive to a native, but the distinctiveness of which cannot or has not quite been put into words or cannot quite be captured even in films and photographs? Is there a way of speaking or manner of expression or a way of pronunciation that we do not know how to document or that can only be "heard" as a result of very long experience with many native speakers? The interrogator might ask the spy to show him how the Semipalatinskians pronounce a certain word—not just tell him, but *teach* him—correcting minor errors as he does it. All teachers know just how hard it is to disguise book-learning for practical experience when confronted with an experienced pupil. These are the ways that the spy will be caught out.[6]

The TEA-laser study referred to above (Collins 1985) showed how laser scientists who had learned their craft solely from printed instructions were equally caught out. In that case they were unmasked when their TEA-lasers failed to work. The general rule is that we know more than we can say, and that we come to know more than we can say because we learn by being socialized, not by being instructed. The unspoken parts of knowledge are a different sort of commodity to the spoken parts: they are of a different substance, they have a different *grammar*. For example, just as these things cannot be deliberately told, even with the best will in the world,[7] neither can they be kept secret from a visitor to the society. It is not possible to imagine the whole population of Semipalatinsk starting to act like Londoners in order to prevent a stranger from picking up their ways of being.

If it is correct, this way of thinking about knowledge has significant implications for the future of intelligent computers; it will not be possible to construct the equivalent of a socialized being by giving a computer explicit instructions. On the other hand, if socially competent machines can be built without the benefit of socialization, social scientists will have to think again; if computers are unsocialized, isolated things, and if knowledge is as social as neo-Wittgensteinian philosophy would have it, then computers ought *not* to be able to become knowledgeable. Something is wrong. This argument applies as much to arithmetic as to spying. How can there be *Machines Who Think*, as the journalists put it (McCorduck 1979), unless they are also "Machines Who Live"—that is, machines who live with us and share our society?

Some optimists believe that machines who think are just around the corner, even though machines who live are still to be found only in science fiction. How can this be if the argument about the social embeddedness of knowledge is valid? How can the argument about the social embeddedness of knowledge be true, with all its implications about the cultural specificity of human behavior, if there can be intelligent machines? The artificial intelligence experiment is, then, not just a problem of engineering or psychology but an empirical test of deep theses in the philosophy of the social sciences. The possibility of a science of mankind, which emulates rationalist models of natural science, is being tested in attempts to build intelligent machines. If there can be machines that act indistinguishably from us, then the philosophical distinctions between action and behavior, and the argument about the peculiar nature of human rule-guided action, will turn out, after all, to be of no significance for the prospects of a *science* of society. The pigeons of philosophy of social science are coming home to roost in the intelligent computer (Collins 1985).

We must explain how it is, in spite of the social nature of knowledge, that we have machines that work. The explanation is not simple. First, machines that work rarely do the same work as humans. I will spend a lot of time showing that this is so. Second, machines that don't work can sometimes look like machines that do work. This is because these machines interact with humans who make good their deficiencies. Third, there is a large aspect of human behavior that mimics machines, and machines can mimic these aspects perfectly. To understand what is going on, we do not need to examine the ever-receding frontiers of computational science, we need only to think hard about machines that already live with us.

Machines That Work

Ironically, the journalists and computer scientists who say that machines that think are just around the corner have been too timid. The paradox is already with us and has been with us for a very long time. To the extent that we are prepared to be puzzled by the next generation of thinking machines we should already be puzzled by some existing computer programs, by pocket calculators, by slide rules, by logarithm tables, and by the printed word. In an important sense, all these are machines that work, and they work in a way just as puzzling as the machines that are more easy to anthropomorphize.

To start at the top, there exist computer programs that, given a spectrum of light, can work out the mixture of chemicals that produced it. Once upon a time, such chemical analyses were pieces of frontier science, and the proper interpretation of spectra must once have been the subject of dispute. Just as in today's contested sciences, this dispute must have been settled by forming social alliances, by forcing people to change their minds, and by breaking their hearts. For a human, learning to interpret chemical spectra still requires an apprenticeship—a period of socialization in chemical society. How can this analytic computer program work, then, without the equivalent socialization?

There are more famous and well-known examples of clever computers. By now almost everyone must have heard of Weizenbaum's ELIZA program (Weizenbaum 1976). Even the Apricot microcomputer that I am using to write this book can cope with ELIZA. For example, it engaged me in the following interchange:

HOW DO YOU DO, PLEASE TELL ME YOUR PROBLEM.
I'm feeling lonely.
IS IT BECAUSE YOU ARE FEELING LONELY THAT YOU CAME TO ME?
Yes, that's right.
YOU SEEM QUITE POSITIVE.
I just want someone to talk to.
WHY DO YOU NEED TO TELL ME YOU JUST WANT SOMEONE TO TALK TO?
Isn't it obvious?
I'M NOT SURE I UNDERSTAND YOU FULLY?
I need someone to care.
WHAT WOULD IT MEAN IF YOU HAD SOMEONE TO CARE?
I would feel happier.
CAN YOU ELABORATE ON THAT?
I need someone.
WHY DO YOU WANT SOMEONE?
etc.

ELIZA is meant to mimic a nondirective psychotherapist. It is a very simple program; nothing more, according to its inventor, than a programmer's parlor trick, yet it has been taken seriously. Weizenbaum relates how his secretary became engaged in a consultation with ELIZA of such personal significance that she refused to allow him to look at the computer screen. A number of psychotherapists, impressed with ELIZA's abilities, proposed that such machines had the potential to replace real doctors at greatly reduced cost. Yet ELIZA had no training in psychotherapy, nor had it the sort of wide socialization that one would expect to be a prerequisite of someone, or something, that could help others through their emotional crises.

Another program that achieved substantial success was Winograd's SHRDLU. SHRDLU appeared to be able to converse sensibly about a "micro world" containing blocks of different shapes and colors that it could manipulate. The world existed only within the program itself, but it was translated into an image on a monitor. SHRDLU could be instructed to move the blocks about, put them on top of each other, and so forth, and then answer questions in ordinary English about their dispositions. The following is a short extract of SHRDLU's performance taken from Winograd and Flores (1986, p. 109):

Pick up a big red block.
OK.
Grasp the pyramid.
I DON'T UNDERSTAND WHICH PYRAMID YOU MEAN.
Find a block which is taller than the one you are holding and put it into the

box.
OK.
. . .
What does the box contain?
THE BLUE PYRAMID AND THE BLUE BLOCK.
What is the pyramid supported by?
THE BOX.
etc.

Once more, this program appears to be holding the sort of conversation that would otherwise only be within the purview of someone who was thoroughly imbued with the culture of solid geometry and the English language.

But the mystery is even more intriguing than these examples suggest. Complicated, clever, state-of-the-art programs make us think that there is something very cunning in the program that enables it to interact with such apparent social accomplishment. The true depth and universality of the problem, however, appears only when we look at much more simple examples. Take a very simple computer—a pocket calculator; this seems to do arithmetic better than anyone I know. Yet doing arithmetic is again a skill learned through socialization and classroom practice. Even this quintessentially intellectual activity is learned within a community; we learn the *language* of mathematics. How can it be that my pocket calculator knows a language when it has never lived outside my pocket? Why stop at calculators? What about my slide rule? There is a sense in which it too can do arithmetic—certainly it and I can do arithmetic together—so it again must be partaking of the language of mathematics; is there a book to be written called *Slide Rules Who Think?* Yet my slide rule is not a social being. Are my logarithm tables social? They speak the language of mathematics with me in the same way as my slide rule.

We need not stop at the language of mathematics. The puzzle of computers, "How is an apparently social activity emulated by a socially isolated artifact?", is the same puzzle as how the printed word can carry knowledge between one person and another. All language is a social activity—how can it be encapsulated in inanimate paper and print? It is interesting that writing was once greeted with the same suspicion as expert systems are now. In Plato's *Phaedrus,* Socrates says:

It shows great folly . . . to suppose that one can transmit or acquire clear and certain knowledge of an art through the medium of writing, or that written words can do more than remind the reader of what he already knows on any given subject. . . . The fact is, Phaedrus, that writing involves a similar disadvantage to painting. The productions of painting look like living beings, but if you ask questions they maintain a solemn silence. The same holds true of written words; you might suppose that they understand what they are saying, but if you ask them what they mean by anything they simply return the same answer over and over again. Besides, once a thing is committed to writing it circulates equally among those who understand the subject and those who have no business with it; a writing cannot distinguish between suitable and unsuitable readers. And if it is ill-treated or unfairly abused it always needs its parent to come to its rescue; it is quite incapable of defending or helping itself. (Hamilton 1973, l. 275)[8]

For Socrates, writing is but a pale shadow of social interaction.

The Social Nature of Artificial Intelligence

We have reached a point whence it is hard to see how to go on. Perhaps the social, enculturational model is wrong. Perhaps, while it is true that socialization is necessary for learning and transfer of knowledge, computers work because knowledge can be stored in a passive form within an isolated machine. There is another way of thinking about knowledge that makes it seem very much the property of the individual rather than the property of the social group. If I lock myself up in a room for a day, so that I have no contact with anyone else, when I come out in the evening my knowledge is not much changed. If it was true that I could speak English but not Chinese in the morning, in the evening I would still be able to speak English but not Chinese. Barring the possibility that I was in some form of extrasensory communication with English-speaking colleagues during the day, it looks as though all that social knowledge was fixed in my head the whole time I was in the room.[9] From this point of view, a facsimile of my head and body constructed during my day of isolation would have all my knowledge without ever being socialized or ever encountering another human being. It looks, then, as though one way to make a perfect intelligent machine would be to take an ordinary human and put him or her through a "Matterfax." This is a device, like a three-dimensional photocopying machine, that replicates the physical structure of matter down to the position of the last electron and the last quantum state. Given the Matterfax, there is nothing in principle to prevent knowledge being transferred to a computer.[10]

From this point of view, the problem of artificial intelligence seems to be about getting the same sort of complexity into a machine as is found in the brain. But this cannot be the problem we are dealing with here. I have argued that the conundrum is essentially the same for computers as for books; they too seem to mimic human linguistic capacity. Therefore it cannot be a matter of complexity. It is not as though a much more complex book will do the trick that a simple book cannot manage, and it is not as though a book mimics the content of the brain. The question we are dealing with is more modest. We want to know how things like books manage as well as they do in their interactions with us given that they are so far in substance and appearance from a Matter-faxed human being. And we want to know if extensions of our current methods of making books, and more intelligent artifacts, will lead us toward the Matterfaxed-style intelligent being by continuous incremental steps that we can foresee. We will not be able to understand how books and pocket calculators do so well by comparing what they do with the content of the brain. A book and a human brain are just too different for this to make any sense. To make progress in this direction we need to ask the question about artificial intelligence in a different way—a way that acknowledges the essentially social features of intelligence. The way that machines, or other simpler artifacts, fit into social interactions should be our starting point.

Fitting in is not always a matter of fitting perfectly. Consider the question: "Can we make an artificial heart?" By that question we mean: "Can we make a heart that will keep someone alive if it replaces his own heart?" The heart is not judged by reference to its own performance but by reference to the performance of the organism in which it is embedded. Suppose we made a heart that was slightly less efficient at pumping blood to the lungs than a real heart, but suppose the body responded to this marginally inefficient implant by producing more red blood cells so that the net amount of oxygen transported by the blood remained the same without any other disadvantages? We would consider this a highly satisfactory artificial heart even though the heart itself did not mimic the original. The same applies in a more minor way to the appearance and composition of artificial hearts. Appearance and composition affect the wavelengths that are reflected when light filters through the body's walls, and they affect the distribution of heat within the chest. But, within limits, the body is indifferent to these marginal changes. An artificial heart that had input and output characteristics marginally different from a real heart, but that was indistinguishable from a real heart in terms of the externally visible working of the human body, would be counted as a fine machine. Thus, from an engineering point of view, the performance of artificial hearts isolated from the context of the human body is not germane.

The same applies to the question of intelligent machines. There are two different questions relating to artificial intelligence. First, there is the psychological question, which is concerned with modeling the processes of the human mind with a computer. For psychologists, the purpose of mimicking human beings with a computer is to learn more about the processes of human cognition. To answer the psychological question we would want to mimic the workings of the brain.[11] What I will call the engineering question of artificial intelligence is quite different. The engineering question is: "Can we mimic the inputs and outputs sufficiently well to keep the organism going, irrespective of whether the mechanism corresponds to the original?"

The crucial difference between an artificial intelligence and an artificial heart is the organisms within which they function. For an artificial intelligence the organism is *not* the human body. When we ask whether we can make an intelligent machine, the big mistake is to think that this is the same question as: "Can we make an artificial brain?" But no one wants to remove a human brain and replace it with something whose artificial nature will not be obvious from the outside (as surgeons want to do with artificial hearts). There are philosophical and psychological debates about whether a person whose brain had been replaced with an artificial substitute with identical inputs and outputs would still be the same person. This sort of debate was once current in the case of artificial hearts, but it is not the sort of question that concerns us here. The organism into which the intelligent computer is supposed to fit is not a human being but a much larger organism: a social group.

The intelligent computer is meant to counterfeit the performance of a whole human being within a social group, not a human being's brain. An artificial intelligence is a "*social prosthesis.*" In the Turing Test the computer takes part in a little social interaction. Again, when we build an expert system it is meant to fit into a social organism where a human fitted before. An ideal expert system would replace an expert, possibly making him or her redundant. It would fit where a real expert once fitted without anyone noticing much difference in the way the corresponding *social group* functions.

Thus in artificial intelligence the question that is equivalent to "Can we make an artificial heart?" is "Can we make an artificial human?" And, just as an artificial heart does not necessarily have to have identical input or output characteristics (including appearance) to a real heart, neither does an artificial human. The embodying organism may be indifferent to variations, or it may compensate for inadequacies. As we will see, this explains the competence of programs such as ELIZA. ELIZA is hopeless as a brain but, in the right social circumstances, acceptable as a human.

The artificial heart analogy can do a little more work before we leave it. The body has an immune system that rejects foreign materials. Other things being equal, to be accepted by the body, an implant has to be designed to fool an alert immune system. There is another approach, however. Luckily for transplant patients the sensitivity of the immune system is not fixed; it can be reduced by drugs. According to the state of the immune system, the same prosthesis might be treated as an alien invasion or as a familiar part of the body. In the same way the social organism can be more or less sensitive to artifacts in its midst; one might say that it is a matter of the alertness of our social immune system. To use a term from debates in social anthropology, it a matter of the extent to which we are charitable to strangeness in other peoples.[12]

The admirable trend in the debates of the 1960s and 1970s was to see things from the other's point of view and thus increase our tolerance and reduce our tendency toward rejection or imperialism. One of the things I will try to do in the last part of the book, however, is to make our social immune system more sensitive to mechanical strangers, for, just as reducing the sensitivity of the physiological immune system carries with it enormous costs for the body, reducing our sensitivity to mechanical invasion has costs too. To learn to recognize artifacts for the strangers they are we need to understand their limitations. This draws us to ask the ultimate engineering question: "Given maximally vigilant humans, not only disinclined to compensate for machine deficiencies but actively seeking them out, what will give the machines' true identity away?" In other words: "What aspects of human ability can't machines mimic?" It is the sort of question that gives point to works of science fiction such as *Invasion of the Body Snatchers*, *The Stepford Wives*, or

Blade Runner, or Isaac Asimov's stories of robots. The possibility of human simulacra is well beyond the scope of this book, but the question is even more philosophically intriguing and informative when asked in more immediately relevant and limited circumstances such as the Turing Test. How can one learn to spot the deficiencies of a machine when communication is restricted to teletype terminals?

One final point of clarification: some people have a principled objection to anyone who says that such and such a technical development is impossible. The future, it is said, cannot be foreseen. In a vacuous sense this is correct. We can think only about *foreseeable* extensions to *current* ideas; it is not a matter of prophesying the future of mankind. I will suggest that certain things are not possible, but I mean by this only that such things cannot be envisaged by extending current thinking—not that such things will never come to pass in unforeseeable futures. Thus, if I say that emulating such and such a human ability is not possible, I will not have taken account of the Matterfax Corporation. I say such and such a thing is impossible in the same way as I might say that it is impossible that we will ever be able to buy a skin cream from the pharmacy that will allow us to take holidays in comfort on the surface of the sun. Perhaps such a thing will come to pass—but not by incremental progress.

2

The Knowledge Barrier and Digitization

The computer is a great success. No one denies the power of the computer as a manipulator of data and as an arithmetician and mathematician. This potency runs from the mainframe through the desktop word processor and database to the pocket calculator. At the fringes of artificial intelligence the manifest successes of certain chess-playing programs, of some elementary robots, and of some simple expert systems show that the computer is as good a logician as a mathematician. With only a few adjustments, however, the limitations of what computers can do still follow the pattern predicted by Hubert Dreyfus as long ago as 1972. Dreyfus set out a fourfold classification of intelligent activities, the fourth of which was, in principle, beyond the capacity of rule-following computers. His argument drew on Wittgenstein's analysis of rule following (e.g., 1953).[1]

> My thesis . . . is that whenever human behavior is analyzed in terms of rules, these rules must always contain a *ceteris paribus* condition, i.e., they apply "everything else being equal," and what "everything else" and "equal" means in any specific situation can never be fully spelled out without a regress. Moreover, this *ceteris paribus* condition is not merely an annoyance which shows that the analysis is not yet complete. . . . Rather the *ceteris paribus* condition points to a background of practices which are the condition of the possibility of all rulelike activity. . . . Thus in the last analysis all intelligibility and all intelligent behavior must be traced back to our sense of what we *are*, which is, according to this argument, necessarily, on pain of regress, something we can never explicitly *know*. (Dreyfus 1979, pp. 56–57)[2]

The Wittgensteinian analysis of rules underlies a large part of my argument too, but the *form-of-life* is the more fundamental concept. The essentially sociological idea of a form-of-life helps us to understand the central importance of the social group and the process of socialization in the way humans follow rules. Although it is true that what we know cannot be reduced to a set of rules, and that attempting to express our culture in terms of rules leads to an infinite regress, and that, insofar as computers require that rules be expressed they will never encapsulate cultural practices, this still leaves a problem: "How does one explain the deficiencies of the various types of computers that do not need all their rule to be made explicit before they are programmed?"[3]

Examples of unexplicit rules include the rules within the hardware of computers; the rules in record-and-playback machines such as those discussed in the next chapter; in chapter 14 a string-searcher is discussed, and this produces close-to-human conversational performance without the rules of conversation needing to be explicitly programmed; machine induction and so-called discovery programs seem to develop their own rules; finally, connectionist machines appear to build their knowledge in sets of connections that are not obviously the equivalent of explicable rules. My *negative* argument is that even though these machines and programs do not appear to use explicable rules they are still not social beings. The fact that rule-following computers cannot imitate us because sets of rules do not encapsulate cultures does not mean that computers that don't use explicit rules are, by that fact, potential members of our society. So long as they are not members of our society, I argue, they cannot imitate our intelligent activities.

The Knowledge Barrier

Another problem with the rules critique taken on its own is that it is too powerful; it becomes a victim of its own universality. If you apply it in a thoroughgoing way, the difficulty is not to explain the failures of computers, but to explain their successes.

One way of coping with the problem is to divide the world of cognition into two basic types of stuff. Thus, in Dreyfus's fourfold classification of intelligent activities the first three types are amenable to computerization (the third, in principle, but not in practice) while the fourth is intractable.

In Dreyfus's scheme the first two types of knowledge domain can be described exhaustively so that meaning and situation are either irrelevant, or meaning is "completely explicit and situation independent" (1979, p. 292). Word-by-word translation with a memorized dictionary fits into the first of these categories. The translation may contain ambiguities, and may not make good sense, but this is the price of a working, mechanical, computerizable task. Such word-by-word translations may help a human translator complete the task of rendering a foreign original into sense even though they are not in themselves examples of competent language use. Fully computable games such as tic-tac-toe, geometry, and other problems that are "completely formalized and completely calculable" fit into the second category. Dreyfus expected (in 1972) to see good progress in these domains and he was, of course, correct.

Dreyfus's third category is in principle the same as the second, but the sheer size of the problem makes it different in practice. Nevertheless, these problems are self-contained, and are independent of external context. Chess is the best example: it is a game that is in principle fully computable, but in practice the number of moves is so great that an exhaustive computational method will not suffice. Effectively this means that the problems within these domains—such as single moves in chess—can be thought of as relating to an "internal context." In chess this is the whole game. Dreyfus expected to see some progress in these domains through the use of rules-of-thumb, or heuristics which approximate the full computational task.

The fourth category of intelligent activities are "dependent on meaning and situation which are not explicit." A good example of the difference between this area and the first is language translation. This area encompasses full natural language translation, whereas the first area covered only word-for-word translation involving a dictionary. This area also covers the remainder of our intelligent activities (Dreyfus 1979, p. 294):

> Since in this area a sense of the global situation is necessary to avoid storing an infinity of facts, it is impossible in principle to use discrete techniques to reproduce directly adult behavior. Even to order the four [areas of knowledge in a table] is misleadingly encouraging, since it suggests that Area IV differs from Area III simply by introducing a further level of complexity, whereas Area IV is of an entirely different order than Area III. Far from being more complex, it is really more primitive, being evolutionarily, ontogenetically, and phenomenologically prior to Areas II and III, just as natural language is prior to mathematics.

The dichotomy between fully exhaustible and open-ended domains is also found in the recent critique by Winograd and Flores (1986, p. 75):

If we begin with the explicit or implicit goal of producing an objective, background-free language for interacting with a computer system, then we must limit our domain to those areas in which the articulation can be complete (for the given purposes). This is possible, but not for the wide range of purposes to which computers are applied. Many of the problems that are popularly attributed to "computerization" are the result of forcing our interactions into the narrow mold provided by a limited formalized domain.

As a means of predicting the domains in which computers would succeed, this type of critique has proved perceptive and lasting. Nevertheless, the dichotomy on which these analyses rest, convenient though it is, is suspect.

The dichotomy is convenient for the critics because it provides a natural barrier to incremental success—what I call "the knowledge barrier." This allows for the existing successes of computers without admitting to their potential for indefinite improvement. There are, however, optimists who, recognizing the more intractable, informal, areas of human ability beyond the domains in which success has been attained, do not see this as an obstacle. The optimists see logic, mathematics, and science as the zenith of human achievements whereas our less formal ways of doing things represent *defects* in our abilities.[4] The steady increase in formal computing power and the steady reduction of human problems to mathematically representable models is, for them, the essence of progress. They think of the world as being made of only one type of knowledge stuff, so that success in the formal sphere promises eventual success in the informal sphere. For the optimists the knowledge barrier is no more a fundamental obstacle than was the sound barrier. As in that case, it is all a matter of more powerful engines. The question is, then, whether there is more than one type of cognitive stuff.[5]

I believe the critics' subdivision of the world of knowledge is fundamentally unsatisfactory for reasons that grow out of their own philosophical grounding. According to the sort of philosophies that the critics take as their starting point—such as phenomenology, or the later philosophy of Wittgenstein—there is no easily attained formal sphere that is fully exhaustible in terms of explicit rules. Perhaps this has been overlooked because of the overwhelming temptation to think of computers as artificial brains rather than artificial members of society. As soon as one thinks of computers as artificial humans the social dimension of even what we think of as formal activity becomes clear, and the magnitude of what has already been achieved by computers becomes a puzzle. The worlds of logic, mathematics, and science are as much human constructs and products of social interaction as natural language. Starting with the same Wittgensteinian ideas one must be more *philosophically* radical than Dreyfus; there is only one kind of cognitive stuff in the world.[6]

Insofar as this is the case, the optimists have the best of the argument. The optimists cannot take too much comfort, however, because the single kind of cognitive stuff *is* what the critics have in mind when they argue for the impotence of computers, and insofar as this is the case, the pessimists have the best of it. The final irony gives comfort to the optimists; because there is no fundamental knowledge barrier, the computer's existing successes must give hope for further progress beyond the spurious discontinuity. Thus philosophical radicalism leads one to a less radical *practical critique* of the potential of computers than that of others who share the philosophical starting point.[7]

The problem is, then, that critiques based on a dichotomy of knowledge types take there to be too much discontinuity between classes of knowledge. This turns attention away from the need for an explanation of what has been achieved on this side of the knowledge barrier; formal problems are made to look too easy and too completely specifiable. Take an area such as arithmetic, in which pocket calculators and computers are acknowledged to be our equals or superiors. Computers seem able to replace arithmeticians in social networks without difficulty. On the one hand, describing arithmetic as belonging to a formal domain does not show how this can be. On the other hand, if arithmetic is a special kind of formal knowledge, different to run-of-the-mill human knowledge, how is it that humans can embrace it with such ease? Is it that in the case of formal acts, pace Wittgenstein, we really do work by fully specifiable rules? How is it that the Wittgensteinian problem of infinite regress in the application of rules does not apply when it comes to arithmetic?

The answer is, as I will go on to argue, that the Wittgensteinian regress applies in principle just as much to arithmetic as to anything else, but that *we* have a special way of coping with it. In this way, *we* have made it easier for machines to mimic us when we do arithmetic than when we speak natural languages. Nevertheless, there is still more to arithmetic than machines can cope with. As I will show in chapters 4 and 5, computers do not do arithmetic as humans do arithmetic, and very often they get the wrong answers—yet they are still excellent partners in our arithmetical world—because we have the ability to repair their interactive deficiencies. Thus a model of social interaction also underlies my *positive* argument. The model shows how computers can and do succeed.

The pocket calculator, then, fits into our social organism where a mathematician once fitted, in the way an artificial heart fits into a human body. Even ELIZA, the principle of whose program certainly belongs in one of the first two Dreyfusian categories (there is no need for any heuristics in ELIZA—everything is mechanical and deterministic), seems to some people to slot quite nicely where an understanding psychiatrist would otherwise fit into their social life. Yet ELIZA, so far as these people are concerned, is fulfilling a Dreyfusian Area IV-type task. The missing link in such cases is that we play a large part in the pocket calculator's arithmetical abilities and an even larger part in ELIZA's success as a psychiatrist. The idea of the knowledge barrier masks this much more revealing continuity.

Digitization

As I have begun to indicate, one of my argumentative strategies is to take existing analyses of knowledge and show how what are usually thought to be properties of knowledge itself are actually differences in the way we attend to it. Thus I have begun to argue that, appearances to the contrary, there is no knowledge barrier and no basic dichotomy within the cognitive stuff of the world; the appearances are to be explained by the way we act. The concept of digitization is another important way of dividing things that can be known into two fundamental types. Is this difference in the knowledge stuff or is it really in us? I will use Haugeland's interesting discussion as the starting point for mine.

Haugeland (1985) says that formal, self-contained systems such as chess and other games are characterized by the fact that the tokens have no *meaning*.

> In some fancy chess sets, *every* piece is unique; each white pawn, for instance, is a little figurine, slightly different from the others. Why then are they all the same type? Because, in any position whatsoever, if you interchanged any two of them, exactly the same moves would still be legal. That is, each of them contributes to an overall position in exactly the same way, namely in the way that pawns do. And that's what makes them all pawns. (Haugeland 1985, p. 52)

Digitization is the method by which the invariance of tokens in such a system is preserved, and digitization is what makes the value of the pieces unambiguous.

A *digital system* is a set of positive and reliable techniques . . . for producing and reidentifying tokens, or configurations of tokens, from some prespecified set of types. (Haugeland 1985, p. 53)

The key concept of digitization, which accounts for the lack of ambiguity, is, according to Haugeland, the margin of tolerance surrounding what is to count as a token of a particular value. Thus a poker chip that is worth ten dollars is still a poker chip worth ten dollars even after it has become worn and scratched. A letter in an alphabet is still the same letter even if it is badly written, partly rubbed out, illuminated, or stylized.

Now, let us depart from Haugeland's analysis and think about the notion of digitization not as a "metaphysical category" (1985, p. 53) but as a process. Consider the value of a gold bar. As it stands, the gold is indefinitely divisible. Imagine we try to cut it into ten equal pieces. To work out the exact value of any piece of gold it would have to be weighed to an arbitrary degree of accuracy. If it were accidentally scraped or chipped after weighing its value might be slightly reduced. Its value is always, at least marginally, ambiguous. Now imagine that the gold bar is cut into ten pieces and each piece is identified by a stamp bearing a symbol that certifies it as one-tenth of the whole. So long as there is confidence in the stamp then each piece can be taken to be worth exactly one-tenth of the whole and each piece is exchangeable with every other piece without the necessity of careful weighing. The value lies in the certification, with its socially accepted margin of tolerance, rather than in the gold itself.

Of course, trouble can arise so long as the certified value is close to the actual value. In these circumstances it is hard to maintain the margin of tolerance and "clipping" will destroy the coinage (see Styles 1980). The solution is to make the tokens out of base metal or paper, in which case the value resides entirely in the stamp and not at all in the material. This reduces all ambiguity to the ambiguity of the symbols, and as we treat symbols digitally this ambiguity is reduced to zero, or near zero.[8]

The tolerance for ambiguity in the symbols arises because we have trained ourselves to recognize them. This, it is important to note, is possible because we can exhaust a universe of *prespecified* symbols, and thus we can learn to recognize every one. Every letter of the alphabet, for example, however written, has to belong to one or another of a fixed set that we have already learned. Tolerance can be maintained when in-between categorizations are disallowed. We can have tolerance for the symbols because they belong to an exhaustive universe—everything must be one prespecified thing or another—whereas the raw gold bar can be subdivided in an indefinite number of ways.[9] Because there are an indefinite number of ways of dividing the gold there is no way of training ourselves to recognize unambiguously the particular lumps into which it might be divided prior to the division.[10]

Thus, the reason that stamped units of currency work is because we have a high degree of trained tolerance for certain symbols that comes with our social conventions specifying an exhaustive set of interpretations. Thus the symbol, coin, or note can be worn or damaged, but we will still count it as worth its full face value. Haugeland seems, however, to read too much into this. He seems not to have noticed the extent to which training and convention play a part in the mutual acceptability of the symbols; he talks as though the digital nature of symbol systems resided in the cognitive stuff itself, rather than our treatment of it:

> But the real importance of digital systems emerges when we turn to more complicated cases. Consider, for a moment, the respective fates of Rembrandt's portraits and Shakespeare's sonnets. Even given the finest care, the paintings are slowly deteriorating; by no means are they the same now as when they were new. The poems, by contrast, may well have been preserved perfectly. . . . We probably have most of them *exactly* the way Shakespeare wrote them—absolutely without flaw. The difference, obviously, is that the alphabet is digital . . . whereas paint colors and textiles are not. (Haugeland 1985, p. 55)

This way of looking at things directs us toward thinking of symbols as having a kind of force upon our minds—a fixedness that is outside of us. The force is nicely represented in the following quotation, attributed to Galileo and used to advertise typeface design (on a poster from Adobe Systems Incorporated):

> But above all astonishing inventions, what loftiness of mind was that of the man who conceived of finding a way to communicate his most recondite thought to whatever other person, though separated from him by the longest intervals of space and time! To speak with those as yet unborn, or to be born perhaps a thousand or even ten thousand years hence! And with what ease! All through various groupings of twenty simple letters on paper.

Here we have the opposite of Socrates' view as expressed in the *Phaedrus* (see chapter 1). In fact, Socrates was largely right, and things are not so simple. It is not symbols, or tokens by themselves that preserve and carry meaning. When we consider the way we interact with symbols and tokens they appear far more ambiguous and far less fixed. Their fixedness is at best a matter of the way we deal with them. To make the most crude of points, the value of a hand of cards depends on the opinion of the man with the gun, and the value of a base metal coin on the opinions of the people who make up the money markets.

To be a little more subtle, let us go back to chess, a paradigmatic example of a formal game using formal tokens and formally prescribed moves that can, supposedly, be fully specified and played in isolation from the outside world. First, consider the game as a whole. When is chess not chess? Suppose that it really were possible to program a computer to play an exhaustively perfect game of chess in the way it is possible to write a perfect program for tic-tac-toe. Would one want to play against it? Would what the computer was doing count as chess at all if you could not possibly win? Or would such a computer be better thought of a chess-practice machine in the way that, say, a mechanical tennis-ball server does not play tennis but, rather, provides an opportunity to improve one's play?

Again, when humans play chess, there are all sorts of subtleties to the game. For example, when I play poor chess with my friends it is not always clear when a move has been made. Sometimes there are arguments about whether the move has been completed—was the finger removed from the piece? Sometimes there are arguments about whether a move is to be counted as a proper move, or not. For example, if one has been playing a long, intricate and interesting game, and then one player, engrossed in a complication, makes an obvious blunder, what does one do? Neither opponent wants an interesting game to end in a simple blunder; often the blunder is retracted and the players can continue with their game of chess, rather than their game of avoid-the-blunder.

The rules of chess, then, vary with the context and content of the game. Even at international level, those who are not partisans first and chess enthusiasts second are concerned with the difference between a game that has been lost by a blunder and a game that has been won by brilliant tactics or strategy. The former isn't really chess; if all chess games ended in unforced errors it would hardly have much of a following. On the other hand, chess played between two perfect computer programs that contained the data from an exhaustive search algorithm would have no following either. It would not even be very interesting to learn from because, presumably, each new game would repeat the previous one. (Nonperfect computer programs for chess, i.e., current and foreseeable programs, are a different matter.)

Now return to Shakespeare's sonnets. I once knew a man whose profession was typography. One topic of his conversation was the extent to which we still had the original Shakespeare preserved. If we didn't have the original manuscript, he felt, then we didn't have the original sonnet. Nuances of punctuation and the actual layout of the verse on the page were part of the original. My first reaction to this was dismissive. This was a man who could look at two pages of, digitally, identical prose and be horrified by one of them while solemnly approving the other. He would point out the uniform gray tone of one page, contrasted with the light and dark patches of the other; the way the *o*'s on one page were exactly level with the other letters, whereas the *o*'s on the other page were, correctly, set a little lower so that they appeared to be level; the way the "risers" on the *h*'s on one page were a different thickness to the risers on the *n*'s and so forth.

At first this seemed like an example of professional pedantry that had no bearing on the meaning of symbols that remained digitally invariant.[11] Then I remembered that my first book (Collins and Pinch 1982) was published at a time when ragged right-hand margins were in vogue. My co-author and I refused to countenance such a "direct edition," insisting that we wanted a proper book with right-hand justification. Our case study was parapsychology and the book needed to look extra serious if the philosophy of science arguments, already radical in themselves, were not to be dismissed out of hand. The publisher appeared to agree, but when my co-author and I received the proofs we found that though it was not literally a direct edition—in that we had not been involved in preparing the pages—the book had been set with a ragged right-hand margin. We offered, to no avail, to forego royalties if the publisher would reset the text. We cared deeply about how the digitized alphabet looked on the page.

There is something of a similar but opposite effect with the spread of laser printers and desktop publishing. It is no longer possible to sift incoming mail by looking at the way it is printed.

What is worse, to make certain that a personal letter can be distinguished from a personalized letter it is necessary to abandon the technology and write by hand. The point is made by the very advertising poster referred to above, which also carries the following apt quotation:

> Typefaces clothe words. And words clothe ideas and information. Clothes, the cliche says, make the person. Hardly. A person, an idea, a word has its own character and its own personality. But clothes can attract or repel, enhance or detract, emphasize or neutralize, and make a person memorable or forgettable. Typefaces can do for words, and through words for ideas and information, what clothes can do for people. It isn't just the hat or tie or suit or dress you wear. It's the way you put it on and the way you coordinate it to your other clothes and its appropriateness to you and to the occasion that make the difference. And so it is with type. A typeface library is a kind of wardrobe with garments for many occasions. You use your judgment and taste to choose and combine them to best dress your words and ideas. (*U&lc*, Vol. 7, No. 2, June 1980)

Thus the contrast that Haugeland draws between the Rembrandt and the Shakespeare is not quite right. The way the sonnets are printed or written does affect their impact—at least marginally. When we spend a banknote, the style of printing or the state of the paper does not affect its monetary meaning but this is because of what we do with the symbols on the note, not because they are symbols. The meaning of symbols does vary when they are used for other purposes such as the writing of sonnets.

When Haugeland says that we still have Shakespeare's sonnets "*exactly* the way Shakespeare wrote them," he cannot mean that they "look" the same as they "looked" to Shakespeare and his contemporaries, or he would not be able to draw the contrast with Rembrandt's paintings. Clearly if we had the manuscripts they would have faded, and so forth. We have just established that the changed appearance of the writing does marginally change the meaning, but the meaning has changed in a more dramatic way too. The meanings of the words in the sonnets has changed as our culture has changed. "Shall I compare thee to a summer's day," just does not mean the same as it did when Shakespeare wrote it. For one thing, the grammar is now archaic, the words are crusted in romance and other associations just because they are the words of Shakespeare, and because they have been spoken so often in so many romantic circumstances. Meanings are not invariant, hence the following joke: "A man takes a fool to see *Hamlet*. At the end of the play he asks, 'Well, how did you like it?' 'Not much,' replies the fool, 'It was just a collection of old quotes.' "[12]

What is it then that is invariant in the Shakespeare sonnet, but varying in the Rembrandt? It could not be the appearance of the thing itself, for just as the Rembrandt fades, so does the writing. Neither does the writing seem to be obviously more easy to reproduce without loss than the painting. Admittedly, many people want to see an original painting, whereas original manuscripts are an esoteric interest. Yet this might just be a conceit. There are reproductions of Rembrandt that people buy and enjoy and, because the original has faded, it is not inconceivable that certain reproductions may look more like the painting as first executed than does the original. What is more, the significance of an original painting cannot lie in its appearance or art lovers would not be so upset when they discover they own a forgery that they had long admired as an original. (In this sense, it is the Rembrandts that are the digitized objects. Irrespective of the way their appearance changes, or relates to other objects, only Rembrandts have high commodity value. There is a tremendous zone of tolerance to their appearance around their high valuation. What counts is not the painting itself but the symbolic certification of originality.)

If the meaning of both painting and sonnet have changed over the years, what is it that makes us say that there is something uniquely unchanging about the words of the manuscript? It is the deeply ingrained and hard-learned, almost reflex-like, ability to see written symbols as the same when we do not have special reason to reflect upon their differences. For example, we are ready to put in a great deal of interpretative *work* to see all those elongated *f*-like symbols as "the same as" our modern *s*'s—to fit them into our prespecified universe of symbols. That is what we are prepared to work at. But it is only that very central part of reading—symbol recognition—to which we have granted the degree of tolerance that makes it a digital system for us. The way we treat symbols is what makes it *possible* to have a digitized system of money, but it is not a sufficient condition as inflationary spirals reveal. Haugeland has mistaken the way we attend to symbols as a property of the symbols themselves, and as a property of that which they are used to represent; it is neither of these.[13]

What Haugeland's discussion of digitization draws attention to is simply this. If we want to discover discontinuities between the decontextualized and the situated, the formal and the informal, the closed system of games and the open system of meaning in the world, then we must look further than the symbols or meanings themselves. The dichotomies are a brute, empirical fact of our being in the world, but their source is not the world, it is us. There are not two sorts of knowledge thing to be found in the world: it is just that there are two ways in which we attend to the one sort of thing.

3

Machine-like Action

Action and Behavior

Dividing up the world of knowledge stuff does not explain why some intelligent tasks can be accomplished by machines and others cannot. There is, however, a traditional way of dividing up the activities of human beings that is a better starting point. This is the distinction between action and behavior. "Acts," as I use the term, are what humans do when they intend to do something, whereas "behavior" is unintended. A standard contrast occurs between a wink and a blink. A wink may be intended, say, as an acknowledgment of a verbal message, a reassuring indication of friendship, or a sexually suggestive advance. A blink is an involuntary movement of the eyelid, not an act or action; it is a mere piece of behavior.

In what follows, I use the term behavior in a slightly extended way to refer not only to unintended movement such as blinks, reflexes, and other involuntary movements, but also to the physical counterpart of action. Thus, in my terminology, it makes sense to say that the *behavioral counterpart* of a wink and a blink are the same—a certain jerky movement of the eyelid; it is just that in the case of the wink the behavior was the consequence of an act, whereas in the other case, no action was involved.

The philosopher Peter Winch (1958), among others, uses this dichotomy to argue against the possibility of a natural science of society. Natural sciences, he argues, deal with causal relationships, whereas acts are related through intentions. What counts as similar pieces of behavior is a matter of agreement among observers, whereas what counts as similar acts is a matter of convention within society, or a form-of-life. In the case of behavior, what is to count as similar is decided outside the object; it is decided within the society of scientists who study the objects in question without reference to the internal state of, say, the molecule, or the pigeon, or the young baby. In the case of action, the actors cannot be ignored because it is *they* and *their* society, not the society of scientists, that establish what counts as similarity of action.

More recently, Searle has used the action-behavior distinction in his critique of artificial intelligence. The point is that computer programs, like putative scientific-style analyses of society, depend on causal-logical relations of events external to us. Computer programs do not partake of the world of intentional action.[1]

Searle (1984, p. 58) provides the following examples of the difference between physical movements and actions:[2]

> If I am going for a walk to Hyde Park, there are a number of other things that are happening during the course of my walk, but their descriptions do not describe my intentional actions, because in acting, what I am doing depends in large part on what I think I am doing. So for example, I am also moving in the general direction of Patagonia, shaking the hair on my head up and down, wearing out my shoes, and moving a lot of air molecules. However, none of these other descriptions seems to get at what is essential about this action, as the action it is.

Searle is not well known for exploring the practical consequences for computers of the difference between action and behavior. Indeed, in his best-known argument—the Chinese room—he starts by hypothesizing the existence of a language-speaking system that speaks perfectly without understanding.

In the Chinese room a person has a code book that instructs him or her how to respond to remarks in Chinese passed through a slot. The code book is a very large reference table that provides a response acceptable to a Chinese speaker for every conceivable input. Searle argues that this reveals the difference between understanding Chinese and merely responding to Chinese symbols. He says that the person in the room does not understand Chinese, and neither does the room as a whole; therefore, even if a machine could be built that would respond adequately in a Turing Test-like situation, it would not prove that it understood language. In my terms, such a language system performs the *behavioral* counterpart of language speaking without performing *acts of speech.* In Searle's thought experiment, action and behavior are indistinguishable to an observer—only the philosophical distinction remains clear.

This kind of "in principle" argument is not satisfying. If the hypothetical machine is *defined* as behaviorally indistinguishable from a person, then the subsequent argument bears on few practical questions. If the machine is indistinguishable, then it must be the social organism's equivalent of a perfect artificial heart—it is bound to be a satisfactory social prosthesis and, however sensitive our social immune system, it will not appear alien.[3] The questions posed here are different. The questions are: "If a machine cannot act but only 'behave,' what difference does this make to the visible performance of the machine when compared to a person?" That is, "Why is the Chinese room an untenable hypothesis?" The Chinese room distracts attention from this question by hypothesizing perfection at the outset.

Machine-like Action

For the purpose of understanding when machines cannot reproduce action and when they can, the crucial differences between action and behavior are the following:

(i) the same piece of behavior may represent many different acts;

(ii) the same act may be executed or (to use computer jargon), "instantiated" by many different behaviors.

The point is simply illustrated in figure 3.1. In this figure, when I wrote the *A* in the second word, and the *H* in the third word, I executed two different intentions with the same behavior (the same movements of the hand). On the other hand, when I wrote the *a* in the first word and the *a* in the second word, I executed the same intention (writing an *a*) with different pieces of behavior.

Figure 3.1
A's and *H*'s

Baker and Hacker (1985, p. 166), in their commentary on Wittgenstein, give the following less contrived examples:

> Punching keys on a computer, enclosing a written slip in an envelope, moving a piece of wood seem very different, but might all be making the same chess move. Conversely a series of acts of writing a name on a slip of paper and putting it in a box look the same, yet might be casting a vote, spoiling a ballot, taking part in a raffle, etc. [4]

The *negative* argument that follows is that a casual-logical machine is not likely to reproduce behavior corresponding to that of an intention-driven human because of the complexity of the relationship between intention and behavior. The link is a matter of social relationships, not causes.

The distinction between action and behavior also can be used, however, to mount a *positive* argument that shows how computers can do what they do. I suggest that there *is* one subset of action in general where the distinction between action and behavior really is of no practical significance to the designer of intelligent machines. My argument is that even though a defining characteristic of actions is that they *may* be carried out in many ways, humans sometimes choose to forgo this option. I define "behavior-specific acts" as follows: *Behavior-specific acts are acts that humans always try to instantiate with the same behavior.*

Though the same behavior is always associated with such an act successfully carried out, behavior-specific acts are still acts in spite of provision (ii), above, because they *could* have been carried out in another way. Part of the act, in such a case, involves *intentionally* not carrying it out with a variety of behavioral instantiations on successive occasions. It is just in the case of behavior-specific acts that the distinction between action and behavior is of no practical significance to the designer of machines intended to mimic humans.

The cliché example of behavior-specific acts is the stylized, Chaplinesque work of people on a production line. The acts are repetitive and identical. F. W. Taylor (1947), the inventor of scientific management, said that in his system the workman would be told precisely what he is to do and how to do it, and any improvement he made upon the instructions given to him would be fatal to success.[5] In this system the ideal production-line worker is supposed to suppress all imaginative and innovatory acts—to suppress everything that would enable an outside observer to distinguish between action and behavior. Ideally, the outside observer would be unable to tell whether it was a human, or a machine disguised as a human, that was performing the task. Thus, on a production line, the act of, say, putting a wheel on a car, normally something that we would do in a variety of different ways—bouncing the wheel to the car, rolling it, lifting with the hands, flicking it up with a foot, fully tightening some nuts before others, or starting all the nuts before tightening them—is ideally done with the uniquely most energy efficient sequence of movements. The preferred description of the act being performed by such an ideal production-line worker would be that he or she was trying to act like a wheel-mounting machine—like a feature of the natural world. Henceforth I will refer to such acts interchangeably as behavior-specific acts or "machine-like acts."[6]

Note the following philosophical points. The idea of a machine-like act is not essentially tied to the idea of a machine. Machines are not part of the definition in a circular or analytic way. Nor are machine-like acts defined operationally. They are not "acts that can be carried out by a machine"; many machine-like acts cannot be carried out by existing machines. The alternative and equivalent usage, behavior-specific act, is less misleading in these respects.

On the other hand, a behavior-specific act is not equivalent to behavior in a number of ways. First, such an act is still intentional; for example, part of the intention might be to act in a restricted way. Second, a behavior-specific act is still a behavior-specific act even when it is not successfully executed in a machine-like way—perhaps because of incompetence or external obstacle. The preferred execution may be machine-like though this may not be obvious to the observer because no repetitious movements are involved. Thus the preferred description of a behavior-specific act is still the intention, not the behavior. Third, it is not possible to tell whether an act is behavior-specific purely by observing behavior even when the behavior is repetitious. For example, one might observe a sequence of identical movements corresponding to someone signing checks, but if the signer knows that the checks for large amounts will "bounce," whereas the smaller ones will not, there are at least two different sorts of acts taking place: defrauding and paying. If, on the other hand, the same movements were being used to show what it is like to "make one's signature," then those repeated movements would indeed be instantiations of a behavior-specific act.

It is important to note that machine-like action is not easy for humans. It takes a lot of training and a substantial effort of will for a human to disguise action in this way. It is also important to note that in spite of F. W. Taylor's ambitions and the popular image of production lines, machine-like action is not widely practiced. On the one hand, the difficulty that humans have in maintaining repetitive performance for long periods may make for marginal inefficiencies in factories. On the other hand, it turns out that most tasks that would seem to be merely routine in fact depend on the operative being an actor rather than a machine, and altering the behaviors used to instantiate the action in subtle ways. For example, in our imagined case of wheel mounting, suppose a chassis arrived with some debris attached to the wheel. A human would remove it before mounting the wheel without giving the matter a moment's thought, but this involves a large departure from the normal, energy efficient, behavioral description of the act.[7]

Another example of behavior-specific action is military drill. The object of drill is to mold the behavior of the men into a machine-like practice. The eventual aim is that it becomes an unthinking routine, but new trainees will try to repeat the drill movements exactly like their fellows in a quite self-conscious way. This may not be efficient or even very effective, but it is an action informed by the intention of behaving like a machine. In subsequent chapters I will look at what happens as machine-like actions become "internalized," but in the meantime note that internalization and unthinking repetition are not the philosophical essence of machine-like action. As in military drill, internalization (that is, unselfconscious performance) may simply be the best way of reproducing the set of behaviors associated with the act.

The difference between behavior and a behavior-specific act is nicely brought out in the fantasy military drill of Joseph Heller's *Catch-22*. In order to win a drill competition the men are effectively changed into machines; their arms are pinned to the same long wooden board so that they are forced to swing in unison without reference to intention. In this case, the behavior is reproduced reliably, but it is clearly not an act.

Machine-like action is nicely caricatured in New York store windows where some mannequins are replaced by humans who act like mannequins. In San Francisco certain street performers stand completely still like statues. Once attention and contributions have been attracted the performer breaks into a jerky dance reminiscent of the mechanical musicians found in fairground steam organs. The performer earns the money because of the extreme difficulty of the mimicry. (Though if the mimicry were perfect no one would give anything since one does not give money to a machine.) A similar caricature of machine-like action is the robot-like dance popular in the mid-1980s. It is a caricature because the movements are not really constrained, just done in a jerky manner suggestive of constraint.

A useful term that makes it easier to see what is meant by saying behavior is the mechanical counterpart of action is "behavioral coordinates of action." I take this term from coordinate geometry. In coordinate geometry one describes a shape—say, a straight line or a curve—by the position of its points with respect to a set of *x-y* axes. Any two-dimensional shape can be described by a set of *x-y* coordinates. One simply takes pairs of numbers (*x*'s and *y*'s), plots them on graph paper (using the coordinates to "drive" a pencil), and the corresponding shape emerges. Although straight lines and curves are familiar enough, the same process will produce a two-dimensional drawing of a face or a cat or anything else, in whatever detail one desires. This process is what underlies the patterns on computer and television screens where electrical signals control magnets that drive the electron beam as though controlled by *x-y* coordinates. In theory, three-dimensional objects could be represented by adding a "*z*-axis," and changes in time could be thought of as represented by a fourth, or time-axis. These are called "space-time coordinates."

Winch (1958, pp. 73–74) asks us to consider the following vivid scenario. He invites us to think of an injured cat:

> We say the cat "writhes about." Suppose I describe his very complex movements in purely mechanical terms, using a set of space-time coordinates. This is, in a sense, a description of what is going on as much as is the statement that the cat is writhing in pain.

One may imagine a mechanical facsimile of the cat designed to be driven by the description of its movements in terms of coordinates. To the outside observer, the mechanical facsimile and the cat would be indistinguishable even though (if one allows a cat intentions) writhing is one thing and the coordinates are another so that Winch (1958, pp. 73–74) is correct to say:

The statement which includes the concept of writhing says something which no statement of the other sort, however detailed, could approximate to.

With "intelligent" robots such as those that respond to feedback from their environment, or those that do calculations of various sorts in determining their next piece of behavior, it is as though there is an additional dimension to the normal four—a kind of logical dimension. With the extra dimension added, the programs of intelligent machines may be thought of as the behavioral coordinates of action. Though the philosophical distinction between action and the behavioral coordinates made by Winch (and by Searle) are valid, the question remains: "Under what circumstances can action be completely substituted by its behavioral coordinates so far as an outside observer is concerned?"

To go back to the production line, quite complex physical movements are already being reproduced by machines like the mechanical cat. Factories use robots that mimic human skills using the technique of "record and playback." For example, such robots can be used to paint chairs.

Consider what Dreyfus (1979, pp. 237–238) (inspired by Wittgenstein) has to say about the problem of recognizing chairs :

What makes an object a *chair* is its function, and what makes possible its role as equipment for sitting is its place in a total practical context. This presupposes certain facts about human beings (fatigue, the way the body bends), and a network of other culturally determined equipment (tables, floors, lamps), and skills (eating, writing, going to conferences, giving lectures, etc). [Can there be context-free features of chairs?] They certainly cannot be legs, back, seat, etc., since these are not context-free characteristics defined apart from chairs which then "cluster" in a chair representation, but rather legs, back, etc. come in all shapes and variety and can only be recognized as *aspects* of already recognized chairs.

Suppose your job was to spray chairs with paint as they passed along a production line. You would have to recognize the chairs if you were not to spray parts of the production line, the conveyor belt, and the supporting hooks, mistaking these for parts of the chair. The problem of recognizing a chair requires a great deal of experience with which no conceivable robot can be equipped. To exemplify, compare modern plastic kits for toy airplanes (see figure 3.2). It is hard to know which bits of plastic are part of the

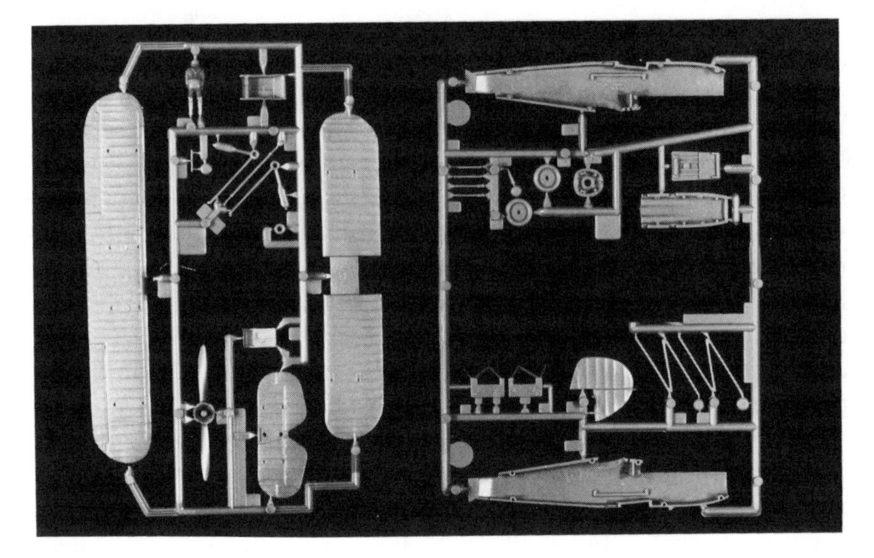

Figure 3.2
Model airplane kit

model itself, and which are parts of the supporting framework for the plastic moldings. It is easy to work out which are the model's wheels, or propellers, because of our everyday familiarity with such objects, but hard to know when it comes to small pieces of body panel, wing struts, and so forth.

Nevertheless, in spite of the philosophical problem of chair recognition, the behavioral coordinates of the act of chair spraying can be stripped from a human chair sprayer and reproduced by a machine. For example, a paint-sprayer head can be fixed to the end of a multijointed robot arm, the movements of which can be monitored and recorded on magnetic tape. The first part of the operation requires a skilled human to guide the spray head through a job, say, the complex maneuvers needed to spray an intricate metal chair. The movements of the spray head are recorded. Henceforth, the spray head can be made to repeat these movements by replaying the tape so that it drives the motors, exactly reproducing the human operator's movements. Provided that the chairs to be sprayed are all the same, so that the operator would naturally spray the next chair, and the next chair, and so on, with the same set of movements, such a device can be used to spray further chairs without the operator. Note that in the record-and-playback method of programming, it is not necessary to *articulate* the craftsman's knowledge, nor even the behavioral coordinates of the movements.

Now imagine that one were to witness such a robot working, and imagine that one did not understand the technology involved. One would see the spray head perform intricately choreographed movements, reaching round, apparently intelligently, to every nook and cranny in the chair, and switching the spray on and off in just such a way as to cover the complex shape with the right depth of paint. The robot is transforming a space-time coordinate description of the human paint sprayer's act of spraying back into movements indistinguishable from the act itself. So long as the robot continued to spray the same type of chair, there would be nothing external to the machine to distinguish the behavioral reproduction from the original act. This machine would appear to be recognizing chairs. We may be able to make an analytic distinction between the act of the human paint sprayer and the behavior of this robot (as in Searle's Chinese room and Winch's cat), but we could not tell which was which just from looking. Thus, so long as things go on without external disturbance, we may say that the act of spraying chairs can be *reproduced without loss by its behavioral coordinates.*

The difference between the human and robot paint sprayer would become obvious, of course, were a chair of a different shape to appear on the production line. The robot would continue to spray in the old way, and its inappropriate *behavior* would spray the paint in all the wrong places, whereas the human would continue as before to perform the *act* of spraying chairs in a competent fashion. This is where Dreyfus's analysis of chair recognition applies once again. Nevertheless, a clever designer could begin to overcome the problem by having the robot trained (by record-and-playback) to spray several different types of chair—the movements being recorded separately—and arranging for it to be able to recognize which type was currently in front of it and to switch to the appropriate section of its program tape.

At first sight this does not seem to resolve Dreyfus's problem, because the robot must still recognize which type of chair is in front of it. But the problem has been changed in a very important and subtle way; the hidden premise is that every new type of chair can be anticipated. Given an exhaustive universe of chair types, each one can be labeled with an easily recognizable symbol. For instance, each type of chair might be labeled with a pattern of reflecting dots—one dot representing chair type 1, two dots representing type 2, and so forth. The robot could shine a light on the label and count the number of reflecting dots. This method of chair recognition is a *digital* method. It rests on the prior exhaustive evaluation of all the symbols and prior "training" of the machine to recognize them and respond to them in a fixed way (see chapter 2). That is the digitization response to the Dreyfusian problem of chair recognition. There is no reason why it should not be extended to a hundred, or a million, different types of chairs—indeed, it can be extended to any number of types of chair that can be anticipated in advance.[8]

Now let us imagine a production line staffed by human paint sprayers but, for argument's sake (perhaps the paint is poisonous and corrosive), they do their spraying by remote control. They work in concealed cabins using articulated linkages to control the spray heads. Now imagine another production line where the spraying was done by record-and-playback robots programmed on the principles described above. Imagine these two production lines side by side in a factory. So long as everything went smoothly, an outside observer would not be able to tell which was which. The behavioral coordinates of chair spraying—executed by the robots—would be indistinguishable from the behavioral manifestation of the act of chair spraying executed by the humans. Nevertheless, the difference would become immediately clear if some unanticipated circumstance not yet fitted in to the exhaustive digital scheme was to occur. As with the example of wheel mounting, one of the chairs might come down the line draped in a piece of debris. In this case, other things being equal, the mechanical sprayer would coat the debris with paint just as if it was painting the chair, whereas the human, in order to execute the act of chair painting, would do something different.

Actually, what I have said in the above paragraph is not quite true because we would expect the line staffed by the humans to exhibit slightly less Chaplinesque precision than the line operated by robots. We would expect to be able to see the difference by the slight, unintended variations in the operations of the humans. But these small variations are what we count as *defects* in what the humans do. They are instances where the humans stray from the Taylorist ideal. The point is that (unanticipated circumstances aside) the factory owner and the operatives would prefer to spray the chairs in the single most efficient way every time. They are engaged in behavior-specific action for they would prefer to instantiate each act of paint spraying with the same behavioral coordinates. A human who was painting many different types of chair this way would be described as engaged in a series of different machine-like acts, each corresponding to a different kind of chair. We can often decompose even quite complex actions into combinations of discrete, behavior-specific acts.[9] It is, of course, just when humans engage in behavior-specific acts that they can be mimicked by machines.

What is Repetitive Action?

What is repetitive action, and why is it easy for it to be mimicked by machines? It may seem strange to ask what is meant by repetitive action. "It's obvious,"one might think. The difficulty is that to say something is repetitive is to say that one thing is the same as another whereas nothing is the same as anything else in every respect. Wittgenstein revealed the importance of this idea. To call one thing the same as another is to have in mind a particular sense of sameness pertaining to a particular form-of-life. This applies even to natural objects. Sociologists of scientific knowledge have shown how much work scientists do in making the world appear constant within their community. For those concerned with this esoteric problem, it is not enough simply to talk of repetitive action as though there were no philosophical and sociological problems associated with the idea.[10]

The answer is that repetitive behavior is specially easy to mechanize because when we talk of acting repetitively we mean that we are trying to act like bits of the natural world. (In the natural world I include everything that is not part of the social world. In my usage the natural world includes machines and other artifacts.[11]) Though there is no metaphysical solution to the problem of what we mean by the same, I want to suggest that our natural attitude is to treat the natural world as though it exhibits sameness. Features of the natural world are, within our culture and practice, our prime examples of constancy. Because we think of features of the natural world as staying essentially the same relative to us, replication of results is one of the basic rules of science.[12] That is why it is easy to model repetitive acts with bits of the physical world such as abacuses, slide rules, phonograph records, Pianola rolls, and computers.

To refer back to the Winchian (1958) point, behavior-specific action is trying to act in a way that makes it appropriate for similarity between acts to be determined by observers without reference to the actors. This makes behavior-specific action, like natural objects, a suitable object for scientific analysis. Science and technology make predictions about the behavior of machines based on external observations. If we make ourselves act in the way we think of machines as acting—"repetitively," as we say—then if our predictions about the behavior of machines are correct, their behavior will correspond to our repetitive actions.

Sometimes machines do not act in the way we predict. For example, machines exhibit more individuality than the idealized model suggests (Kusterer 1978). On the other hand, it is just when we fail in our ambitions to perform behavior-specific acts, as in failures to act out the Taylorist ideals, that reliable machines are preferred to humans. In these circumstances, not only do the behavioral coordinates reproduce the act without loss, but we count reproduction of behavioral coordinates as an improvement over our actions.[13]

The link between digitization and behavior-specific acts is that the intended behavior that corresponds to a behavior-specific act is fixed and could be represented by a symbol. Indeed, one might think of the reflecting dots on the chairs as representations of the discrete, decomposed sets of chair spraying coordinates of action. Even though a human might not be able to execute the coordinates exactly, the actual movements are best described as an approximation to an ideal—the idealized set of movements represented by the digital symbol. There is simply no equivalent for other types of act, because there is nothing to relate the many behaviors that can correspond to ordinary acts to one another except the social rule that is being followed. The classification of ordinary acts is a matter of local social convention not a matter of science. In the case of machine-like acts, the corresponding behavior is digitized and is available for scientific description and analysis, and technological reproduction.

Past and Future

A characteristic of machine-like action is that it can be described without loss (as far as an outside observer is concerned) in terms of its behavioral coordinates. This is a property that is shared with actions that have already been executed. That is, it is a property that is shared with the past. Descriptions of action in terms of behavioral coordinates can be accomplished *after the event* without loss. Completed acts have, as it were, already left their trace upon the multidimensional graph paper of the world and the trace cannot be surprised by what is still to come. The past has already been fully anticipated.

That both machine-like acts and retrospectively considered acts share this property leads to a great deal of confusion. Because the world can be described retrospectively in a simple way, it is assumed that past descriptions put us in a position to predict the future by extrapolation. Our argument shows that this is not the case because, machine-like action aside, even if *action* continues predictably, future *behavior* is not related in a simple way to past behavior.

Suchman (1987) explores the problem when she compares models of human cognition based on the planning metaphor with human action, which she sees as situated. Suchman draws a distinction between formal description and the endless complexity of situated action. Cognitive scientists, she claims, are under the impression that human action is *governed* by plans, because it can be described retrospectively as *following* plans.[14] A plan cannot cope with varying situations in the real world, however.

Plans, then, are like rules. They do not contain plans for how to apply them in particular instances, whereas human action is characterized by the way it deals flexibly with each new and unpredicted situation. The pervasiveness of the planning metaphor for human action is due, of course, to the possibility of describing action as conforming to a plan *after the action is over*, in just the same way as action can be brought under a rule *post hoc* even though the rule itself cannot fully account for the action.

> The fact that we can always perform a *post hoc* analysis of situated action that will make it appear to have followed a rational plan says more about the nature of our analyses than it does about our situated actions. (Suchman, 1987, p. 53)[15]

Thus, actions that do not follow (explicable) rules may still be brought under a rule in retrospect.

The confusion between the analysis of the past and the future accounts for the confidence that some psychologists and computer scientists have in their abilities to describe human action in terms of formulae. It accounts for the computer hacker's confidence that each new modification to a program that deals with a newly discovered exception to a rule (i.e., adding a new rule) will be the last modification necessary. It also accounts for the shattered dreams of a thoroughly scientific sociology and a rationalist philosophy of science. The confidence is misplaced and the dreams have been shattered because success in retrospectively bringing something under a rule does not mean that the rule will account for future, unanticipated, instances.[16]

It is as though scientific sociologists, theorists of history, rationalist philosophers, computer scientists, hackers, and the rest see a truck being driven through a featureless desert. They see the tracks on the desert floor and notice that they provide a complete description of where the truck has been. Some notice regular characteristics in the tracks such as the regular indentations caused by the treads and proclaim that this explains the progress of the truck. They are like logicians, or cognitive scientists concerned with the fine structure of human reason.[17] Others notice more general patterns in the overall shape of the track and by careful analysis fit the shape to something like a mathematical series and try to predict its next few terms. They are like people with a theory of history, or a scientific model of social progress, or a philosophical model of the development of science. Others appreciate the complexity of the track and endlessly refine the model with new terms. They are the econometricians and computer hackers. Look inside the truck, however, and you will see that its progress is governed by a gang of cutthroats, each fighting for control of the steering wheel. The truck is history, society, reason, science, culture.[18]

This metaphor applies *unless* the action being observed is machine-like. In that case, and only in that case, there is complete symmetry between retrospectively bringing action under a rule and anticipation of future behavior. In behavior-specific action we drive the truck in a regular pattern. Thus the tracks in the desert really can be used to predict the progress of the truck, and pace Suchman, there is no distinction between rules and situated actions. That is why behavior-specific action can be reproduced by machines with programs based on what would normally be an oversimplified version of human action expressed in plans or rules.

What Next?

I have argued that the idea of machine-like action, as exemplified in chair spraying, does not preclude enormous complexity.of behavior. Chairs might come in a million or more different shapes, all of which require that any machine following the behavioral coordinates associated with chair spraying switch to the corresponding set. These sets of coordinates will spray the million different shapes without loss compared to human action. That accepted, the questions, to which I continually refer, remain unanswered: "To what extent can we approach a complete description of action by thinking of it as broken down into many small behavior-specific acts?" "To what extent can the ramification of *behavior* approximate human *action* in general?" This might be said to the be subject of the whole book. I will anticipate to this extent: nonmachine-like action could never be distinguished from behavior (that is, the behavioral coordinates of action could completely reproduce nonmachine-like action) only if the set of behavioral coordinates were infinitely long. But much less is required to mimic the larger part of action, and much, much, less to produce something satisfactory for the many occasions when humans fill in the gaps.

4

Mental Acts and Mental Behavior

Behavior-specific acts, or machine-like acts, restrict the instantiations of action to one fixed set of behavioral coordinates. I now want to argue that this idea can be applied just as much to things that go on in the mind as to things that are done with the hands. This argument is not much to do with psychology, nor is it about what happens in the brain; so far as I know, it has little to do with neurons and synapses. It is an abstract description of aspects of cognition, but one that corresponds to everyday observations of the way we carry out what I call *mental acts.*

To establish the connection between mental acts and production lines in a gradual way, think about how what we do with our minds is related to what we do with our bodies. Consider first how one might use one's fingers to add. To use fingers in this way, one must already know how to count—that is, one must be able to chant the names of the natural numbers in order. Here is a transcription of my young daughter's written account of the procedures involved in adding four to five using her fingers.

> *First* you say "5." Then you put up 4 fingers [by counting them out] and then you say, "Well I started with 5." And then you count the fingers starting from 6. "Six–seven–eight–nine."

The four selected fingers are used to monitor the recitation of number names beyond five so that it continues for four, and only four, further places in the sequence.

How children learn to recite the names of the natural numbers in sequence, given the big, booming, buzzing confusion of sound that surrounds them, I do not know, but let us say that they have learned.[1] What this means is that, ideally, all recitations of parts of the number sequence are done in a machine-like way—they are *chanted.* Admittedly, children chant numbers with different intonations and rhythms at different times but these are irrelevant to the act of counting. The essence of the act—the intention that identifies it—is that it is the same performance every time.

In the next part of the adding exercise there are two basic ways of going on (starting at four and adding five fingers' worth of places, or starting at five and adding four fingers' worth as described above), and each of these ways may be done with either the left or right hand, with more or less vigor, counting in a loud or soft voice, and so forth. Again, however, the central part of doing the sum is still machine-like. Someone following a coordinate description of the behavior associated with addition using the fingers would be indistinguishable from someone *performing the act* of addition using fingers. The act, in other words, can be described without loss by describing only the behavior that corresponds to it. It is, metaphorically as well as literally, a digital procedure.

Notice that there is nothing very cognitive involved in adding in this way—nothing more cognitive than is involved in, say, spray painting a chair. The essential procedures are chanting, while raising fingers, and chanting again while holding the fingers in the air. It is a very mechanical-looking procedure, and it is taught by *rote.*[2]

Now, at some later stage, a person learning to count will abandon the fingers. Perhaps the next stage is just to imagine the fingers and chant the names of them as they are raised and lowered in "the mind's eye." At a later stage still, chanting aloud might be replaced by whispering, then by lip movements alone (after the fashion of those who have difficulty reading silently). Later still even the lip movements will go and there will be no externally visible behavior that corresponds to the act of counting. There is no point during this sequence of stages when adding becomes something essentially *cognitive.* Nothing special happens to the nature of the act as the externally visible manifestations drop away. Adding in this way is a behavior-specific act that happens to have no externally visible behavior corresponding to it. I call such an act a mental machine-like act but, because I argue that there is no metaphysically significant difference between this kind of act and one that does have visible behavior associated with it, I will drop the qualification "mental" whenever convenient.

There are, of course, further stages in learning arithmetic. As adding becomes fast and automatic, the procedures become *internalized.* At this stage there is no need to imagine any fingers in the mind's eye or to do any silent counting. Adding can be done "without thinking" by someone who is good at mental arithmetic. This does not make the act especially cognitive either, it just makes it more efficient in the case of the human organism. When soldiers learn to drill, they eventually cease to think about what they are doing; in the same way, when children learn to add they cease to think about what they are doing. In the case of the soldiers we are not tempted to regard efficient drill as something cognitive; this is because the behavior associated with the drill remains visible. Addition is not different in kind.

Military drill is, on my account, a series of machine-like acts; adding in the way my daughter describes is similar. Both can be reconstructed without loss by acting out their behavioral coordinates. This is obviously true of drill and equally, if less obviously, true of addition and other mathematical procedures (see chapter 6, especially the section on using a slide rule). If I had forgotten how to add, I could recapture the ability by following rules like those explained by my daughter (so long as the ability to count, the ability to recognize a finger, the ability to remember when a finger has been used once, and so forth, are taken for granted).[3]

Of course, most of our mental activity is not machine-like. If we accept the idea of mental acts, then most of what we do when we use our minds is the mental equivalent of ordinary action with all the associated variation and unpredictability of mental behavior. To go back to the earlier example, think of the different ways Shakespeare's sonnets, or the play *Hamlet*, can be appreciated. Nevertheless, a central part of what we do in the way of calculating, reading, and communicating with one another is usefully thought of in terms of the mental equivalent of machine-like action. Our response to symbols of all kinds, including the symbols of which Shakespeare's works are made up, is to be understood as mental behavior-specific action. The discussion of digitization in chapter 2, which showed that the value of symbols rests on our readiness to respond to them in a fixed way, can be understood in this light.[4]

To switch back to something that is more readily understood as action, let us invent an imaginary teaching regime. Let us imagine that we are first taught the letters of the alphabet by being made to type them on typewriter keyboard. Each time a letter is presented to us, whether it is printed in a book, or written in handwriting, or its name is spoken aloud, we have to recognize it and then respond by pressing the corresponding key on the keyboard. We are said to know the alphabet when we press the appropriate key every time. When we master that ability we are able to execute certain behavior-specific actions, pressing the keys in response to the presentation of symbols. Just as on the parade ground we execute a response to the command "Right wheel" in a fixed way every time, so we are able to execute a "response to the letter tee" in a fixed way, every time. Now to use the same argumentative trick as we used for addition, imagine that once the keyboard has been mastered it is taken away and we reveal our competence in responding to presentation of letters by pointing to where the keys had been while speaking the names of the letters aloud. The next stage is merely to utter the names without pointing, following which we merely move our lips when shown a letter; complete mastery requires no obvious physical counterpart to the response at all. Note that we might eventually manage all this without thinking about it at all (that is the normal and most efficient mode of complete mastery of a skill), or we might do it by attending carefully to what we do. That distinction is not a crucial point of principle. The point of principle is the fixedness of the "mental behavior" that instantiates the mental action. Internalized or not, it is the mental equivalent of Taylorism.

In our earlier discussion of Shakespearean manuscripts we saw the process at work in our determination to respond to elongated *f*-like symbols and the rest of Shakespeare's writing in the same way as we respond to modern printed symbols. The *symbols* produce in us—because we want it to be that way—a fixed, machine-like response. Thus a variety of different typefaces or handwritings are all *seen* as the same. A part of our potential for acting on variation is nipped in the bud. To put ourselves in a position to begin to disagree about the meaning of words we must volunteer to live a bit of our lives in greater interpretative poverty. By ignoring one potential area of variability *we* digitize the symbols in words. We turn them into tokens as far as we are able—tokens that, like currency, have mutually agreed fixed values and can be exchanged among ourselves. That is what formalization comprises: the general agreement to give up a bit of *individual* human potential, which would lead to mutual incomprehensibility, for the sake of the beginnings of mutual understanding and much more productive misunderstanding.

To switch again into a more practical metaphor, learning to read is like learning a musical instrument. To master the piano, the hand must be trained to act like a tool and just the same applies to the eye and mind as it learns the code for writing.[5] But bear in mind what has been mastered when the hand has completed its training at the piano—completed it even to the extent that all its skill is no longer consciously accomplished but "automatic," or "reflexlike." At that stage of piano playing one has learned to play the notes in sequence, but one has not learned to understand music. One has not even learned to understand music as "a set of old quotes."

Machine-like action involves acting in ways indistinguishable from following a set of space-time-behavior coordinates. These coordinates fit smoothly into the causal-mathematical-logical description of the world. A human engaged in successful behavior-specific action looks like an automaton propelled by physical causes. Once we have internalized the skills this is even how it feels to us.

Once the skill is internalized we seem to relate to signs in the same way as we relate to obstinate features of the natural world that provide causal constraints on our physical movements. As we stumble against a rock we do not seem to have to think about obeying its instructions. It will give us guidance about where to walk in its vicinity whether we think about it or not. We may walk beside it or away from it, but not through it. We do not have to *decide* not to walk through it. Our actions are caused directly by the rock rather than by our interpretations of what the rock is. The same applies to a reflex. We do not decide to blink when something moves suddenly toward our eye—the movement causes us to blink. Once internalized our relationship to sign systems feels like this.

And yet, in other respects, the effect of these signs is not like that of a rock face. First, they only affect us in this obstinate way if, at some time, we were ready to be affected. They only work within cultural frameworks. A rock instructs everyone equally and obtains uniform results. Rocks seem to be cultural universals[6] whereas the symbols of mathematics affect only those who have accepted a mathematical training, and the symbols of the alphabet affect only the literate. The symbols of English affect only those who know the language—those who have gone through the training.[7]

The second contrast is that although rocks command a response without needing to be recognized (after all, one might bump into a rock at night), responding to symbols requires that they be recognized first.[8] Thus, before we can even begin to undergo a program of training in machine-like responses we must be able to solve the problem of induction. In this case the problem arises in its pattern recognition variant: "When is one letter shape to be counted as the same as another?" Haugeland points out that an *a* is treated as an *a* even if it partly rubbed out, but he does not explain how a partly rubbed out *a* is *recognized* as an *a*. The notion of tolerance just does not account for it. Only *after* we have solved the pattern recognition problem can we exercise the tolerance necessary to support a digital symbol system—tolerance is not a sufficient condition of similarity of judgments. The point is illustrated by figure 3.1: we need to recognize the *a*'s in the two upper words as *a*'s before we tolerate their dissimilarity, whereas the symbols in the two lower words have to be recognized as different *in spite of their similarity*. To describe our responses as simply "causal" is to beg the question. All we can say is there is an unsolved problem, one that applies to humans in the same way that it applies to computers.[9] A miracle of inductive inference takes place before we get to the training stage, which in turn precedes the causal-reflex stage.

It might be thought that recognition and response are the same thing. After all, in my imaginary typewriter-keyboard training program it would be impossible to separate success in letter recognition and success in making a consistent response to the recognized letters. Most of our learning is like this. Nevertheless, the two processes are more than analytically distinct. Imagine a world in which all letters of the alphabet were always written in the form of machine-readable characters (as on checkbooks), were always the same size, always printed in the same quality ink, and only ever presented to the eye at the same distance and angle and in the same lighting conditions. The problem of recognition would be substantially reduced. For example, in such a world machines would be as good at recognizing letters as we are. Nevertheless, the question of response would remain the same. A person undergoing the typewriter-keyboard training regime, would still have to learn the proper responses to the letters. It would still be possible to respond by noticing a familiar splotch of ink rather than the letter as a letter. Imagine that some foreign symbols, such as the Hebrew letters *alef, bet,* and *gimel,* were accidentally mixed up with the English alphabet. The students would learn to recognize them without learning any special response. Alternatively, think of the students learning to recognize and respond to the English capital letter *P*. They have reached the stage where they have learned to make the sounds in response to the letters. But the teacher has taught them Greek as well as English responses. As they are presented with a series of letters they normally make a *p* sound whenever they see a *P*. Sometimes, however, the teacher will shout out "Greek!" and they will respond to the same letters, recognized in the same way, with a *r* sound.

The problem of interpretation also would remain in a world of machine-readable script. *Hamlet* written this way could still be a profound experience or a collection of old quotes.

The nature of the pattern recognition problem can be nicely illustrated by writing in unfamiliar script or by mirror writing. Figure 4.1 is a piece of mirror writing. Without trying to read it, note its pattern of repeated shapes and quickly try to make a copy. Now hold the original and your copy in front of a mirror. This will reveal whether you have managed to copy the writing without understanding its meaning. Copying mirror writing gives some insight into the problem of pattern recognition as faced by a computer.

Interestingly, this difficulty is used by forgers to improve their craft. When forgers want to copy some script they first turn it upside down so the meaning is lost. That way they concentrate on the shape of the line alone, without being disturbed by preconceived notions of what an *a* or an *h*, or whatever, should look like. In our terminology, the forgers "undigitize" the symbol system so that they reduce their tolerance to variation in the lettering; this helps them avoid a stereotypical response and attend more easily to the variation in pattern, which is what they want to repeat exactly.[10]

A third contrast is this. Machine-like modes of action cause our behavior only to the extent that we continue to allow them to do so. When reading text, we are able to switch to, say, the typographical designer's way of seeing things or wonder at the ambiguity of signs. What is more, we can use a nonmachine-like way of going on to improve over what can be achieved with machine-like action alone. My description of reading is a caricature of skilled reading, where recognition of individual letters hardly features.We have already imagined the effect of debris on a production line and how we might react to it. In arithmetic exactly the same is true. The arithmetician is always looking for imaginative shortcuts to save

Figure 4.1
Strange writing

going through the full machine-like repertoire. "Five, add four, take away four" is best done without adding and taking away at all. No fingers, or their mental equivalent, are required for the good mathematician to do this sum. As I will show, computers are poor arithmeticians when it comes to competition at this nonmachine-like level.

Right at the center of our mental life there is, then, a set of actions that we perform in a machine-like way. We always instantiate, say, certain acts of symbol interpretation with the same (mental) behavioral response. In the case of the paint sprayer we found that it was behavior-specific action that could be mimicked by machines. In the same way, it is the mental equivalent of behavior-specific action that computers can take over from us. The computer cannot handle the inductive processes that come prior to these reflexes—for example, it cannot handle symbol recognition, nor can it handle the acts of understanding that come after the reflexes, nor can it handle the unanticipated variations of behavior that human action occasionally manifests in even the most routine-seeming tasks.[11] But it can handle the middle bit just as well as we do, and usually better. The partial answer to the paradox of the thinking computer is the human being who, from the outside, might just as well not be thinking. Machines can emulate all those elements of human action that are machine-like or behavior-specific. That is how it is that in spite of the Wittgensteinian-Dreyfusian critique, my pocket calculator seems to do arithmetic.

The Mathematical Room

If the argument of the previous sections is correct, then it must be that pocket calculators and mainframes alike cannot really do arithmetic at all, only a small mechanical bit that happens in the middle of arithmetic, whereas the humans who use the calculator do all the rest. It must be that humans do all the arithmetical acts that come logically prior to and logically after the mechanical part. We can see that this is so by elaborating Searle's Chinese room thought experiment in a rather different way.

Imagine that nearly all our computers and calculators have been destroyed in a catastrophe. To do complex calculations we now go to a special room—the "mathematical room." Outside the room are a number of booths housing numerical keypads of the sort found on a pocket calculator. The keypads are connected to wires that lead into the mathematical room. We go to a booth and use the keys to type out the numbers and symbols corresponding to our desired calculation; as we type the symbols are shown on a liquid crystal display. We finish typing the symbols. After a delay, the answer appears on the display. Apart from the delay, it is just as though we had done the calculation on our own calculator; the inputs and outputs of the mathematical room are identical to a calculator's inputs and outputs. In fact the calculation has been done inside the mathematical room. What might be inside the room?

To begin with, let us imagine that inside the room is just one of the few remaining calculators in the world in the charge of a "specialist." The numbers and symbols that appear on my display are passed into the room by telegraphy and appear on another display next to the specialist's calculator. The specialist types out the symbols on his calculator; he types them one by one in the order that they appear. As he taps in the last key-press the answer appears on his calculator display. The specialist transcribes the digits onto another keypad that transfers the number telegraphically to my liquid crystal display outside the room.

Now, the specialist need not have done any arithmetic for the mathematical room to work. He need not have understood anything of the symbols, the significance of their interrelationship, or the meaning of the digits in the answer. All he needs is to be able to distinguish between the symbols and recognize their sequence. To say that what the specialist was doing was arithmetic would be like saying that every time I wiggle the slider on my slide rule I am doing arithmetic (see chapter 5).

What I have established so far is that the mathematical room can do all the arithmetic that a calculator can do without anyone inside the room understanding anything about arithmetic. It does not get us very far because, since the arithmetic was done by a calculator inside the room, we have merely pushed the problem from outside to inside. Let us, then, experiment with other possibilities for what might be inside the room.

Leave everything as before but substitute a large look-up table for the specialist's calculator. The reference table would have a left-hand column containing all possible sequences of inputs (this will be a finite list because there are only sixteen symbols on the calculator key-pad, and we will set a limit to the number of symbols in sequence that the mathematical room is prepared to handle), and a right-hand column with the answers. Does this make a difference? The answer must be no because the specialist could still refer to the look-up table without having any idea that he was doing arithmetic. Therefore, if it was the calculator that was doing arithmetic before, it must now be the table.

What if the specialist had a slide rule instead of a look-up table? Once more one can see that he could follow a sequence of instructions for manipulating the rule and producing answers without having to know that he was doing arithmetic; he would just have to be able to recognize numbers and follow some rules (which are, incidentally, explained in chapter 5). The same applies to an abacus and to logarithm tables; the specialist could have learned how to use these aids by rote without understanding the nature of calculation. What if the specialist used a pencil and paper and did the sums in the way we learned in elementary school with all the painfully acquired rules about "borrowing" numbers from this or that column, "carrying" them, or "storing" extra numbers above or below the line and so forth? Still, the same applies, just as it applies to doing sums on one's fingers.[12] What if the specialist did everything in his head by mental arithmetic but still produced identical answers to the calculator? On my account even this would not necessarily be any more arithmetical; it would just be that the visible behavior associated with doing the sums would have dropped away.

The mathematical room can accomplish everything that the calculator can do yet the specialist is not doing arithmetic. Where, then, is the arithmetic going on? The answer is that arithmetic is not going on, at least, it is not going on in the mathematical room nor was it going on in the calculator.

Is it that the specialist needs to understand what he is doing rather than just going through the motions; would that make what he does into arithmetic? It depends on what is meant by "understand." The most noteworthy attempt to understand arithmetic is Russell and Whitehead's *Principia Mathematica,* but even Russell and Whitehead, on their own admission, did not really understand it in the end. That is the wrong kind of understanding. There is, however, another kind of understanding.

Consider this sum: '"If I am 5 feet 9 inches tall, and there are 2.54 centimeters to the inch, how tall am I in centimeters?" The answer, according to my calculator, and according to the mathematical room, is 175.26 centimeters (69x2.54). But, if someone says to me, how tall are you in centimeters, I do not reply "one-hundred-and-seventy-five-point-two-six centimeters," I reply, "one-hundred-and-seventy-five centimeters." The unapproximated figure is not the answer to the question. To give the right answer, one must know the use of the calculation within the rest of social existence. It involves knowing the social agreements about what constitutes a correct calculation. Thus it involves knowing how exact to make a calculation, knowing when to give a rough answer even if a more exact one is available, and knowing whether, and how, to make a rough calculation when it is not possible to make an exact one even if an exact one would be better. It involves knowing when a slide rule, a calculator, an abacus, logarithm tables, and pencil and paper are counted as doing the same thing and when they are not. For example, when I was at school, we could use slide rules in our physics tests, but were instructed to write "(s.r.)" after the result of every slide-rule calculation; this was to allow the examiner to discount a level of inaccuracy in the last significant figure (or accept an untidy, approximate answer to a problem that should have resulted in a simple "round number"solution). Knowing how to calculate included knowing that an inaccurate result with (s.r.) written after it was the same as a more exact result without the (s.r.).[13]

If I knew something of numbers but nothing of heights, being told that someone's height was "175.26 centimeters" would still have an effect upon me. It would have the same effect as being told that, say, the rest mass of the neutrino is "4eV." Being told the "4" part of 4eV is like having someone swing their hand toward my eye. The latter will make me blink, the former will engender some equivalent mental behavior. As far as this mental behavior is concerned, the effect of the figure "4" in the phrase "your dining table is 4 feet long," is precisely the same as the effect of the figure 4 in the phrase "the neutrino has a rest mass of 4eV," yet I know a lot about the world of dining tables and nothing of the world of eV's (electron-volts). The difference is that with my new knowledge I can do something more in the world of tables—for example, I can work out whether I can seat eight for dinner—whereas I can do nothing new in the world of neutrinos.

Though I now know that the neutrino has a mass of 4eV, the only thing I could do with it would be the same kind of thing as the specialist in the mathematical room does with the numbers passed in to him. I can use the phrase in some sort of non-eV-related act—such as writing this chapter, or showing off at cocktail parties, but

I cannot use it in acts where it would make the slightest difference if the mass were 20,000 instead of 4 (or if it referred to the mass of a neutron, or even a neuron, rather than a neutrino). The second figure has precisely the same value to me as the first as far as all these are concerned, even though the difference between 4 and 20,000 is well known to me in the sense that it triggers a different response, and in the sense that it would affect my table-related acts enormously if it referred to the length of dinner tables in feet instead of the rest mass neutrinos in eV's. I could learn lots more facts along the lines of the neutrino having a mass of 4eV, and I could even do some manipulations with the data such as working out that the weight of a dozen resting neutrinos plus half a neutrino is exactly 50eV's. These manipulations would pass for arithmetic in school, and certainly pass for arithmetic when they are done by my calculator or the mathematical room, but they are not the sort of arithmetic that is of use to anyone outside these areas of machine-like action. The 50eV result is just like the 175.26 centimeter result for my height in these respects.

One might say that the things I can do with my knowledge of the rest mass of the neutrino relates to neutrinos in the way that cargo cultists' airfields relate to airplanes. The cargo cultists of the Pacific Islands discovered, during the Second World War, that airfields brought airplanes that brought quantities of consumable cargo. When the war ended and the airplanes went away, the cargo cultists built their own airfields and decoy airplanes in an attempt to attract the cargo back again. Cargo cultists use what they know about the shapes of airplanes and runways to engage in religious acts, not in aeronautical acts. Their knowledge vis-à-vis aeronautics is rote knowledge. They know just a very few of the behavioral coordinates of flying.

Now we understand why my calculator, even though it mimics the central aspects of the performance of arithmeticians is only doing those aspects of the act of calculation that can be done in a machine-like way; the calculator deals with numbers in the way that I deal with numbers when, say, I am calculating the rest masses of collections of neutrinos. When I do this I am drawing only on those central parts of my arithmetical skill that I normally execute with fixed behavioral responses. The calculator is reproducing the behavioral coordinates of the act of calculation—the *skeleton* of the act—and when I use it, I am doing the rest. In the mathematical room I am the only thing that is doing what we normally mean by arithmetic. The extent to which parts of arithmetical acts can be delegated is the extent to which we ourselves could substitute parts of the act by a set of behavioral coordinates; this is also the extent to which we could learn to perform the role corresponding to those aspects of the act by rote.[14]

Note how the mathematical room differs from the Chinese room. The Chinese room performs identically to a Chinese speaker. The mathematical room performs identically to a *mechanical calculator*, not an arithmetician. I use the example not to show that when machines do arithmetic they have no understanding but to show precisely the opposite—that machines, because they have no understanding, cannot do arithmetic. This is not a matter of in-principle distinctions; the room actually gives the wrong answer to the question about my height in centimeters. More details of the arithmetical failings of calculators and computers in general will follow in chapter 5.[15]

Same Behavior, Many Acts

The feature of the action-behavior distinction not so far discussed is that the same piece of behavior can be the visible counterpart of many different acts. To go back to the mathematical room, the specialist, in executing a calculation, may have been merely following the behavioral coordinates of the act of arithmetic, but he may not. Suppose I want to know how tall I am in centimeters, and I pass the calculation "69x2.54" to the mathematical room. It may just happen that the specialist wants to know how far his car will travel if driven at 69 miles per hour for 2.54 hours, in which case, while he executes the behavioral coordinates for my act of calculation, he may actually be doing a quite different *act of arithmetic* himself (to which the *correct* answer is exactly 175.26 miles). On the other hand, such a calculation might be incorporated into quite a different sort of act. One might go through the motions of arithmetic after the fashion of arithmetical cargo cultists, so that the intention that provides the preferred description of the act is a religious one. But, because the part of arithmetic we are describing is machine-like, and because the behavioral coordinates for this part of arithmetic can replace the act without loss, what would be praying for the cargo cultists could still be arithmetic for someone else. Likewise, readers of this book might want to multiply 69 by 2.54 for reasons of their own. In that case they can use the behavioral coordinates of arithmetic that I have used to make a philosophical point in their own sum for their own quite different purposes.

Children provide real-life examples of the varied potential of calculators. My son and his school friends use their calculators for races. One may have quite a satisfactory race by keying "1x1.1=" and then pressing "=" repeatedly. The first person to reach 100 on the display panel is the winner. Does one want to say that in this case the calculator is doing arithmetic?

Here is another example: try the sum 34 + 5,771 = 5,805. Now repeat the sum with the calculator upside down. You will find that "he" added to "ills" equals "sobs"! (See figure 4.2 and invert the page.) Is the calculator doing arithmetic under these circumstances? The answer is that it does not make sense to ask what the calculator is doing—it always does the behavioral coordinates of arithmetic. It only makes sense to ask what the kids are doing. The kids are doing acts of something other than arithmetic even though their finger movements on the keys are the same as would be required for the arithmetic itself.

The Mathematical Room, Digitization, and Induction

I have used the mathematical room to explore the inside of my pocket calculator—not its electronic or logical operations, but the human actions to which it is equivalent. We have discovered that the calculator doesn't do arithmetic after all; it only makes a contribution. This contribution is often thought of as the formal part of arithmetic, the part that lies on this side of the knowledge barrier. I have argued that the formal part of arithmetic is not something that exists independently in the knowledge stuff of the world, but is made by our always instantiating certain acts with the same behavior—mental or physical. The formal aspects of arithme-

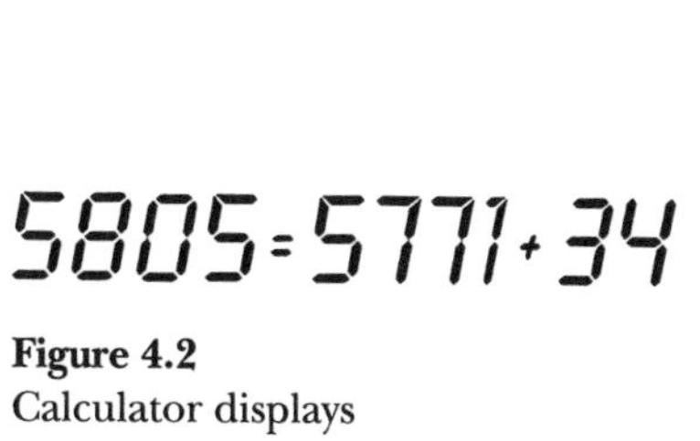

2.54

5805=5771+34

Figure 4.2
Calculator displays

tic lie within a sandwich of human interpretation. When we do arithmetic, part of what we do has to be done through machine-like actions; it is only this part that the calculator takes over from us.

Humans live in societies in which what is to count as similarity and difference is continually being created, renegotiated, established, and maintained. Existing similarity relations are not fully specifiable, though they can be brought under a rule post hoc. These ex post facto rules, however, are subservient to the relations themselves. We simply recognize *examples* of similarity and difference—we know when people are going on in the same way and when they are not, even when two similar acts are instantiated by different and unforeseen external appearances (behaviors). We cannot explain how we do this: we just happen to have these inductive abilities. These abilities are what enable us to live, in the normal way, in a world of *concerted action.*[16] There is, however, a subset of our acts that have a different kind of similarity. In these, the external behaviors are specific to the act. In the case of these acts it is only what counts as similarity of *behavior* that needs to be established and maintained. This can be done without reference to intention. When we do these behavior-specific acts, we live in a world of *concerted behavior* indistinguishable from the natural world. *Digitization turns concerted action into concerted behavior.*

Machines can live with us in a world of concerted behavior, but they do not have the inductive propensities to live in our world of concerted action. When machines partake in that world their inputs first pass through a sieve of digitization. The input devices of computers (including various aspects of the program) are digitization sieves. This is obviously the case with keyboards, but any pattern-recognition or template-matching program (such as is used in voice recognition) has the same function. To understand our interactions with any kind of computer, whatever its internal construction or programming, think first about how its inputs are digitized.[17] The output of computers is also digitized. It fits naturally into the world of concerted behavior but to fit with the world of concerted action we must interpret it and occasionally "repair" it (see chapter 5). These processes are represented schematically in figure 4.3.

The theory of behavior-specific action explains why we can have pocket calculators that seem to join in our social life—the life of the language of mathematics. Just the same applies to slide rules, look-up tables, and, on a much broader canvas to the alphabet. That is

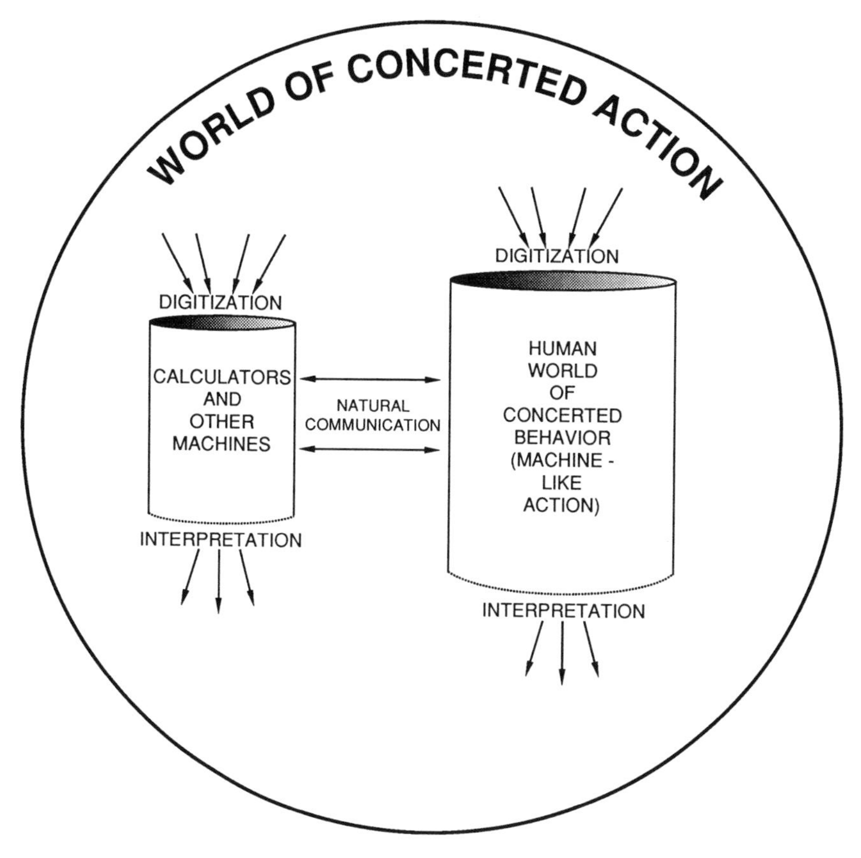

Figure 4.3
Concerted action and concerted behavior

why the written page can figure in the social life of language, but also why Socrates was right to distrust it. The written page does not figure in the same way as the act of speech. The written page is merely the behavioral counterpart of the act of speech, but a behavioral counterpart that corresponds to a central feature of the human use of language. That new machines such as computers do not seem strange to us, no stranger than books or slide rules, is because the world of concerted behavior has long been part of our lives.

5

Interpretation and Repair

Using a Calculator

The theory of machine-like action explains how machines such as pocket calculators fit into that part of our society that I have called the world of concerted behavior. It is we who make possible the passage between *concerted behavior* and *concerted action.* We do this by our interpretation and repair of machines' inputs and outputs but most of what we do goes unnoticed. Consider the process further. What actually happens when one uses a machine such as a calculator? I now describe in detail how I used my calculator to convert my height from inches to centimeters:

First I had to know about the existence of different units of measurement and the principles of converting one to another. Then I had to find the conversion ratio. I was able to look this up on the back of my old slide rule where it was printed for easy reference. This told me "Inch=25.4mms." I know that "mms" stands for millimeters and that there are ten millimeters to a centimeter. I converted this in my head to centimeters by dividing by ten (move the decimal point one place to the left.) Next I did a sum in my head to convert my height into inches. I know that there are 12 inches to a foot. "Five twelves are sixty," I repeated under my breath with the same machine-like rhythm I learned nearly forty years ago, and I added 9 to make 69 inches. Now I knew the numbers were 69 and 2.54 and, using some other mathematical skill, I knew they had to be multiplied together. I switched on the calculator in order to enter 69. My familiarity with keyboards and with calculators told me that I could do this by keying a 6 followed by a 9. The keyboard gave me no choices regarding textual styles. However I visualized the numbers, I was restricted to using the symbols on the keys; my choices were restricted to a set of ten digital alternatives with nothing in between.

Even the way of entering the numbers was once not so obvious as it seems now; it is quite unlike writing or even typing. In writing, the 6 would be entered in the left-hand column to leave room for the 9 to be written subsequently. In typing, the carriage and paper move along to the left to leave room for the 9 to be written as soon as the 6 has been typed. But the calculator writes the 6 in the right-hand column and it stays there. It looks as though there is no space for the 9 to the right of the 6; it will either overwrite the 6 or appear to the left of it. Actually the 6 will pick itself up and jump along to the left, but only when the 9 is entered. I can just about remember when this seemed a very odd thing indeed. I believe I can remember, just for a moment, wondering how to enter a two-digit number on my first calculator.

I say that I see a 6 appear on the screen, but in fact it is a very stylized figure. To have seen that I had entered 69, I must have executed my reflex responses to the numbers on the keys, and the stylized numbers on the screen, but in addition I must have performed the regular miracle of induction involved in seeing a stylized sixty-nine that looks the same as a more regular 69 (much as in figure 4.2). I had to reinterpret the output of the screen to fit it into the world of concerted action.

Now I wanted to multiply that 69 by 2.54. I keyed an ×. Disturbingly (that is, it would have been disturbing, once upon a time), nothing at all changed on my calculator display. I trusted it, however, and keyed 2. Everything disappeared except the 2! I had to trust the machine to have stored all the rest. Experience told me everything was OK. I now needed to enter the decimal part of 2.54 but looking at the display I noticed that the decimal point (a little square) was already in place. I had not noticed this when I entered the 9 to go with the 6 to make 69 or I would have been worried about whether entering the 9 would make 6.9 rather than 69. Making a guess—there was little else to do—I pressed the key with the dot on it. Nothing happened! Again, hoping for the best, I keyed 5 and the little square shape did indeed pick itself up and jump along with the 2 so that it reappeared between the 2 and the 5. Quite unlike ordinary writing or typing, pressing the little square shape does nothing immediately, it merely instructs the decimal point to jump along when the *next* key is pressed.

Am I making too much of the skills required to enter computations into a calculator? No, I am not. Consider the experience of a certain innumerate student at this university learning to use a calculator for the first time. He had to calculate the position of a point on a regression line by multiplying a distance along the *x*—axis (352) by the slope (–.84104). Quite reasonably for a new calculator user he entered the following sequence of symbols:

3 5 2 x – . 8 4 1 0 4 =

He obtained the answer 351.15896. Crucially, he did not have enough arithmetical skill to see that the answer was wrong. He could not, therefore, learn to repair the calculator's mistakes.

To return to *my* story, I keyed the 4 to make 2.54. Any moment now in this story I am going to describe how I took advantage of my calculator's power.

I pressed "=." The answer 175.26 appeared on the screen. The calculator had finished its job.

I considered the display. I performed another unconscious miracle of inductive generalization and saw the stylized liquid crystal display as digital symbols—numbers to which I responded in a machine-like way. The next bit is quite complicated; I did a quick check on the calculation to make sure it made sense and that there was a reasonable chance that I had keyed in the numbers correctly and controlled the calculator correctly. To do this, I already had to know roughly how tall I am in centimeters.[1] I know roughly what a meter rule looks like—it is about a yard long—and I visualized such a rule stood up against me, with a second rule on top of it. This would be about six feet and I am slightly shorter than six feet: therefore, I thought, I should be slightly shorter than two meter rules, or slightly less than 200 centimeters. The result, "175 centimeters" was in the right area, but seemed a bit short. "However," I thought, "a meter is actually a bit more than a yard, so it's probably OK."

The calculator did not do very much of the calculation, just a little bit in the middle. I had to do a whole lot of work on the input side before the calculator did its job, drawing on my broader knowledge, experience, and other skills, and I had to do a whole lot of work on the output side after the calculator did the calculation, drawing on the same sort of commonsense abilities.

Apart from knowing how to use the calculator as a specific type of machine (how to enter numbers and so forth), pretty well all of this knowledge and ability was widespread in the population of potential calculator users long before calculators were invented. They had already been developed in the use of arithmetical and nonarithmetical devices such as slide rules, log tables, abacuses, pencils and paper, and typewriters. One of the reasons we tend to think the calculator can do arithmetic is the natural way we help it out and rectify its deficiencies *without noticing*. All the abilities *we* bring to the calculation—everything that surrounds what the calculator does itself—are so widespread and familiar in their regular mediation between the world of action in general and the world of concerted behavior that they have disappeared for us, only to be made visible by painstakingly deliberate attention. These abilities are now just part of our background common sense.[2]

Machines, Mistakes, and Repairs

We all make mistakes. That computers make mistakes does not, therefore, support the argument that they are fundamentally less competent than us. Nevertheless, computer mistakes are not, in general, like our mistakes. The mistakes computers make reveal that they mimic our machine-like action, not our regular action. Consider spelling mistakes: a word processing spelling checker is most unlikely to make a mistake over a complicated word, but it cannot correct a wrongly spelled word that matches its dictionary. For example, if I type "I love my weird processor" and use the spelling checker, the program tells me "weird processor is spelled correctly." It complains, however, about "antidisestablishmentare-anism" (but not "antidisestablishmentarianism"). Compare this with a child or only moderately literate adult. Both would notice the "weird processor" as an unlikely usage in English, but the long word would almost certainly defeat them. Template-matching speech transcribers would have similar quirks. They would be very good at recognizing complicated words, but poor at recognizing familiar aural ambiguities ("gray tapes"—"great apes")—precisely the opposite problem to that of the human audio-typist.[3]

Something similar applies to the arithmetical mistakes made by machines. I have dealt with one type of mistake—such as mistakes of approximation brought about by the failure of computers to be embedded in the social and cognitive world pertaining to what the particular calculation is about—but computers make another kind of mistake in their arithmetic that few of us ever notice. Mathematicians aside, we are so certain that computers are infallible at arithmetic and so blindly charitable in subliminally repairing their errors that it is hard to convince people that they actually do make simple mistakes. They make these mistakes because even the heartland of arithmetic is not completely machine-like. When we learn arithmetic we learn what are called "simplifications" and "shortcuts"—that is, we learn some *departures* from behavior-specific routines. Actually these are vital if we are to obtain what we count as the right answers to our sums.

For example, if I use my pocket calculator to do the sum 7 divided by 11, multiplied by 11, it produces the answer 6.9999996—a typically calculatorish mistake. In the same way, 10/3x3 produces 9.9999999. These problems arise from doing the sums strictly according to the sequence in which they are presented. Some nonmachine-like action is needed to sort out the *order* in which to do the calculation if it is to come out right. The pocket calculator cannot manage it. But, just like the paint sprayer, the calculator can be improved.

On my desk I have a more substantial calculator that gives correct answers to the problems that fooled my pocket calculator, and is, in a sense, less distinguishable from a human. This big calculator is particularly intriguing: 10/3 gives 3.333333333, and if this is multiplied by 3 in a continuous process, then the answer comes out, as 10. But, if you start again and enter 3. followed by as many 3's as you can, and then multiply by 3, the answer comes out as 9.99999999, not 10. It seems to be able to distinguish between 3.3 recurring as representing 10/3, and 3.3 recurring as many times as it has room for on the display—this is quite annoyingly intelligent. To see the difference between it and a human arithmetician requires more ingenuity. On this machine, if I ask it for 10 divided by 3, store the answer in memory, then multiply the answer by itself, then multiply by 3 again and 3 again—that is, (10/3)x(10/3)x3x3—to which any reasonable arithmetician will give the answer 100 without any trouble at all—the machine gives 99.99999999. (The same machine gives 100 as the answer to 100/3/3x3x3. It has too little ability to do that part of the act of arithmetic that involves "seeing" that this sum and the last are *the same*.) Thus, just as in the imaginary paint sprayer, the more complex design at first hides the difference between the act of arithmetic and its behavioral coordinates, but more vigilant observation reveals the difference. We can go on.

My microcomputer gives the correct answer (100) to the same problem expressed in BASIC (A=10/3; B=A*A; C=B*3*3; PRINT C), but the wrong answer (999.9999) if we extend the sum to just one more term—(10/3)x(10/3)x(10/3)x3x3x3—(expressed in BASIC as A=10/3; B=A*A*A; C=B*3*3*3; PRINT C). Thus, though my microcomputer cannot do simple arithmetic either, it requires somewhat more vigilance to discover this. For a while, it seemed as good at arithmetic as I am.

My son's new "scientific" calculator gave the right answer to the last sum (the intermediate 10/3 being stored in memory and multiplied by itself three times), but could not do "3 to the power 1/20 to the power 20"—that is, the answer came out as 2.99 recurring. More worrying (especially for someone learning mathematics) is that the reciprocal of 1.2x.8333333333 turns out to be 1, and the reciprocal of that 1 is .99 recurring, whereas the reciprocal of any other 1 entered directly is 1. (The problem is that if you don't know where the 1 on the display comes from you might think that the reciprocal of 1 varies from time to time.)

To continue, my university's mainframe at first handled all the above problems. A little more vigilance, however, revealed a number of interesting deficiencies. For example, given the sum 7/11x11 (expressed in BASIC as A=7/11; B=A*11; print B), it at first gave the correct answer 7. But the program has the ability to work more accurately in what is called "double precision." The same sum, done *more accurately* by the computer in double precision mode, produces the answer 6.99999988079071. In this example, what is more accurate for the arithmetician is less accurate for the computer, and vice versa. Of course, even in double precision, the computer was well able to handle the problem if the multiplication was done first, showing that in double precision mode it did not have the capacity to see 7x11/11 and 7/11x11 as the *same* sum. (If we wanted to be theoretical about this, we might say that the law of commutativity had broken down at this point, but the notion of recognition of sameness links the problem to the more general philosophical point.)

A similar problem could be found with the sum 3 to the power 10, to the power 1/10 (expressed in BASIC as A=(3^10)^.1; PRINT A). This comes out as 3 in double precision, whereas "3 to the power 1/10, to the power 10" comes out to 3.00000047683716. A slightly more complex version of this sum—3 to the power 10,000, to the power 1/10,000, which even a moderate arithmetician can see is, in a sense, *the same* as the previous example, defeats the computer entirely. The numbers are too big for it to handle, and it complains of an overflow error. These examples show that the mainframe is not really doing even the "formal" part of arithmetic, but only something that is close to arithmetic.

The above examples might be said to show that even what we normally think of as the formal parts of arithmetic are heavily convention bound. There is a sense in which 6.99999988079071 is the more correct answer to the sum 7/11x11 than 7, but it is not the more correct answer for most arithmeticians. The correct answer depends on what one means by the correct answer. It is tempting to think that the formal part of the arithmetic has a logic like that of an apparently straightforward arithmetical series such as 2, 4, 6, 8, where the continuation 10, 12 seems an inevitable consequence of the numbers themselves. In fact, it has a logic more like that of the series 19, 13, 20, 23, 20, where the next two terms are 6 and 19. The numbers in this second series are the positions in the alphabet of the initial letters of the days of the week expressed in English. The series depends on linguistic conventions. Linguistic conventions are more easily recognized as conventions because natural languages vary from place to place. My argument is that the series 2, 4, 6, 8, 10, 12 also depends on the conventions of a language—the language of mathematics. It is just that the language of mathematics is much more uniformly spoken so that we do not notice its conventional nature.[4]

Our determination, in the normal way, to repair the computer's faulty arithmetical interactions can be nicely brought out by the examples of computer error discussed in the above paragraphs. One is extremely unlikely even to notice that the computer is making this kind of mistake—professional arithmeticians aside, one would never expect that a mainframe computer would make such a simple error. Even I did not expect that the mainframe was unable to handle 7/11x11 and it was some time before I tried it out. My first attempts to fool it were with very high powers, such as $(3^{10,000})^{1/10,000}$ when it complained of a single precision overflow. Pushing my experiment along, I tried the same sum with double precision, and then various other sums with single and double precision. The results were at first very confusing and inconsistent; the computer would sometimes give the correct answer whereas at other times it appeared not to be able to do the very same sum. Looking for order in the confusion I tried working the sums through algebraically without the computer needing to ask for intermediate numerical results (for example, A=3; N=100; B=(A^N)^(1/N); PRINT B'). Getting the right result to this kind of thing, I concluded that the computer had an algebraic processor within its version of BASIC and could do the algebra that would save it from unnecessary evaluations. Only then did I stumble upon the confounding variable of single and double precision. It was not a matter of algebraic form, it was just that I had asked my algebraic question in single precision and my more numerical question in double precision. My respect for the mainframe, cynic though I am, had made me seek a much more sophisticated explanation. It was only after a lot of experimenting of this sort that I even thought of going back to trying 7/11x11. I was genuinely surprised at the outcome in spite of what I imagined was my very uncharitable attitude.

The way that repairs work also can be explored by showing these computer deficiencies to people who use mainframes for routine arithmetical work. It would be interesting to collect a catalog of explanations that are used to save the computer's arithmetical reputation. My embryonic catalog includes: "It's a problem with our mainframe but probably not with another machine"; "You are asking the computer to do a different problem from the one you set yourself. You try doing each step separately—you try to work out 7/11 and then multiply it by 11 and you will get the same answer"; "If you express it so that the indices are multiplied out first, as $(3^{1/10})^{10} = 3^{(1/10 \times 10)}$, you will get the right answer." Most of these claims are true, of course, but presented in this way they simply disguise the fact that the computer cannot do what we can do quite easily. In each case the blame is shifted from the computer to the user; the computer's deficiencies are ignored and all the stress is put on the user's failure to repair the deficiency in what has become, for most of us, the natural and expected fashion.

Of course, *now that they have been mentioned,* it would not be hard to write little "hacks" that would cope with these particular deficiencies so as to save us the trouble of repair. A little routine could be written for the computer to be able to spot when things could be canceled without having to evaluate them—at least when they were given in the form of the sums just described. Each case has left its behavioral trace and future occurrences of the same case could be catered for in a behavior-specific way. This is just to reiterate the point about the asymmetrical nature of retrospective and prospective analysis—the tracks in the desert. My thesis is, however, that such a process could never put the computer in a position to anticipate all, as yet unforeseen, opportunities to cancel (or otherwise simplify) in more complex formulae—this is the job of mathematicians. Mathematicians will always be seeking and finding exact solutions to problems that are currently done by approximate, numerical methods. The process is parallel to seeing the trivial simplifications described above.

On the other hand, it is true that the increasing competence of the machines I describe makes it harder and harder to spot their deficiencies at arithmetic when compared to humans. Given marginally "intelligent" programs that could, for instance, recognize the word "tall," it might even be possible to invent a procedure that would enable a computer to approximate correctly when asked for my height in centimeters. This kind of extension of computers' abilities is the subject of chapter 7.

A Final Metaphor: The Dancing Machine

What we have done when we have learned writing or arithmetic is to learn a dance. Like the dance of the bees it is concerted behavior. We have learned the same dance as many other people who inhabit a great domain. Enter a ballroom and there are an indefinite number of ways of walking around the floor. It is overwhelmingly unlikely that the steps you take and the steps of others will match even though everyone understands what it is to walk properly. (They understand the act of walking but it allows lots of freedom in the behavioral instantiation.) When you dance in a formal way, however, you volunteer to give up all but a restricted and stylized set of movements. These movements are set out in books explaining ballroom dancing. They show you where to put the feet, using diagrams—a white outline for the left foot and a black shape for the right foot. Naturally, in real dancing there is a quite a bit of residual freedom, which is why some people dance with flair and some do not, but let us think of an idealized version of the dance-learning process.

Because others have learned the same dance—have volunteered to suffer the same restrictions on their freedom—your steps and theirs will coincide. In contrast to walking, there is only one way to waltz. The waltz amounts to restricting the act of foot movement to execution through just a limited set of space-time coordinates.[5] That is how others' acts of dancing come to coincide with yours. Insofar as the waltz is thought of as a set of fixed steps (rather than an act of civility, romance, or whatever) we could easily have a machine waltz for us. We could use a waltzing machine as a tool when we wanted to waltz with somebody but felt too tired to manage it ourselves.[6] "May I have the pleasure of this dance?" "Yes." "My machine will see to it." If both parties felt too tired they might delegate the whole process to their two machines, concentrating all their energies into looking into each others' eyes.

That is the extent to which humans can delegate their acts to machines. We cannot delegate *acts,* we can only delegate the behavioral coordinates of the act, and we can delegate these successfully just to the extent that part of the act is sufficiently stylized to be reducible to one behavior and describable without loss by a formula. The *steps* of the dance will coincide with the steps of many others only if this behavioral restriction is widely accepted. The same applies whether we talk of physical activity like dancing, or the production of standardized artifacts, or mental activity such as arithmetic. That is why we can delegate parts of arithmetic to pocket calculators but also why, when we use the calculator, we must still do the equivalent of gazing into our dancing partner's eyes.

The apparent force of the symbols of mathematics arises out of the long history and wide spread of this kind of stylization. The mental act has become stylized across such a great region of the universe that it appears to be independent of us. When we internalize these routine acts, they even *feel* independent of us. Compare this with a case where the restriction of freedoms, far from being internalized, has not been agreed in the first place. In lands remote from Western culture the calculator is no more an arithmetician than two sticks rubbed together are a slide rule.

II

Expert Systems and the Articulation of Knowledge

6

What We Can Say about What We Know

I used the theory of machine-like action in Part I to explain the extent to which computers, including pocket calculators, slide rules, and record-and-playback paint sprayers, can replace humans in social life. Many such machines do not require that knowledge be articulated before it can be mimicked. In this section I narrow the focus to a particular class of intelligent machines that deliberately take our *explicable* knowledge as their foundation. Expert systems reproduce only the articulated behavioral coordinates of action; they do this by extracting and encoding the knowledge of experts as the basis for their programs. In the next five chapters I examine the idea of expert systems both to explore their potential and limitations, and as a foil for more extended analysis of the nature of knowledge in general.

Expert Systems

An expert system is a computer program. Most expert systems differ from other programs in that they encode relatively familiar knowledge expressed in nonmathematical form. The programming languages of many expert systems are designed to reflect the ordinariness of the knowledge base, and programs can be comparatively easily read by the nonexpert. Figure 6.1 is an extract from an expert system for semiconductor crystal growers written with "AUGMENTED PROLOG FOR EXPERT SYSTEMS"—APES (Collins, Green, and Draper 1985).

These specialist programming languages can make decoding and alteration of expert systems easier, and they may also make it easy to query the knowledge base of the system. There is no reason why such systems have to be about technical subjects—they can easily encapsulate rules of thumb ("heuristics," in the jargon) that would not normally be formalized in this way. For the purpose of

```
X wont-be-dissolved-by-or-reduce (Y|Z) if
        not (Y|Z) dissolves-container-material X and
        not (Y|Z) will-be-reduced-by X
(X|Y) dissolves-container-material Z if
        (X|Y) contains-metal and
        Z is-a-metal
(X|Y) will-be-reduced-by Z if
        (X|Y) contains-an-oxide and
        Z is-a-reducing-agent
pyrex will-soften-at 501
quartz will-soften-at 1251
```

Figure 6.1
Extract from an expert system for crystal growers

```
X will-come-back-to-my-place if
        X has-a-fleeting-smile and
        X will-maintain-prolonged-eye-contact and
        not X is-short-sighted
```

Figure 6.2
A simple expert system: PICKUP

illustration, I wrote a very small program in APES called PICKUP. The complete program is represented in figure 6.2. It could be thought of as a kind of miniature expert system. An actual consultation with PICKUP is represented below. (Read ":" as "such that")

find (X:X will-go-back-to-my-place)
WHICH (X:X HAS-A-FLEETING-SMILE)
ANSWER IS Mary
ANSWER IS Delilah
ANSWER IS Albert
ANSWER IS end
IS IT TRUE THAT MARY-WILL-MAINTAIN-PROLONGED-EYE-CONTACT?
no
IS IT TRUE THAT DELILAH WILL-MAINTAIN-PROLONGED-EYE-CONTACT?
yes
IS IT TRUE THAT DELILAH IS-SHORT-SIGHTED?
yes
IS IT TRUE THAT ALBERT WILL-MAINTAIN-PROLONGED-EYE-CONTACT?
yes
IS IT TRUE THAT ALBERT-IS-SHORT-SIGHTED?
no

==> ALBERT

As can be seen, the reasoning works in a kind of tree structure.[1] From the initial query the ramifying branches of the tree are explored, touching here and there on the user's knowledge until a satisfactory answer that makes sense in the inquirer's universe is reached. But, as we will see, the question of how and where the tree touches the user's universe is a vexed one.

In some expert systems (such as those written in APES) the program makes it possible to save the information "learned" at each consultation. If the above consultation had been saved, then the next time PICKUP was asked the same question, it would immediately conclude with the answer "Albert" without the intervening queries. Thus the programs appear to learn by the experience of interacting with their users. After a knowledgeable user had spent some time with the program, it would "know" quite a bit more than it did at the start, and subsequent users would feel the benefit of this new learning. This seems to make the programs still more like experts.[2]

The specialist languages, useful though they are, are not the essence of expert systems. Any computer program can, in principle, be written in any computer language, so the programs listed above *could* have been written in, say, BASIC, albeit with more difficulty; all computer languages amount to strings of 0's and 1's (or on/off switches, or whatever) when they are translated for action in the processing heart of the computer. There is no agreed definition of an expert system; some think that the essence is the language, others that a probabilistic inferencing mechanism is the vital ingredient, and so forth.[3] The definition of expert systems I adopt has two parts:

(i) An expert system is a computer program designed to replace experts in social interactions.

This part of the definition on its own is too broad; even my pocket calculator has the potential to replace experts in social interactions. Thus there must be a second part to the definition:

(ii) Expert systems are largely based on the knowledge of human experts.

(To save complications, in thinking about the social interactions in which systems are to replace experts, I will discuss only programs that interact through language.)[4]

"Knowledge elicitation" is the term used to describe the process of extracting knowledge from an expert for subsequent encoding, whereas people who write the programs for expert systems are often referred to as "knowledge engineers." As with spying in Semipalatinsk, the question is the relationships between our competences, what we can learn by explicit instructions, what we can then say about what we know, and how this relates to what an expert could say. But experts engage in different types of verbal interaction. One may present the expert with a problem and ask: "What should I do? Give me an answer." Here, the expert is being used as a consultant, and the means by which the problem is solved is not germane to the questioner. If a computer is used in this kind of role, the only reason why its program should be based on human expertise is if that is what works best.

A quite different kind of interaction is involved when the expert is used as a teacher, for example, an apprentice master.[5] To be a teacher, an expert system has to have a capacity to explain, which is problematic enough (see chapter 7), but more fundamentally, it must have the ability to solve problems in the same way that humans solve them. Any way of solving problems will be good enough for an expert consultant, but only the human method of solving problems will serve a machine that is to take the place of a human teacher. This requirement again shows the importance of the second part of the definition.

Knowledge, Rules, and the Acquisition of Skills

Internalizing Explicit Rules

The way I experience my knowledge has something to do with the knowledge itself, and something to do with the way I have been taught to think about it. Perhaps because of our educational systems, we tend to think of the paradigm case of knowledge as a mixture of information and logico-mathematical ability that can be recapitulated in an examination. It is as though we think of all our knowledge as being fully exhausted by what we could say about it—fully exhausted by a set of behaviors, including writing, which correspond to it. The relationship between what I can express and what I know is, however, much more complex.

A straightforward theory is the three-stage model of learning (Fitts 1964; Anderson 1982). Berry (1987, p. 7) presents this model:

In stage 1 (cognitive stage) an individual learns from instruction or observation what actions are appropriate in which circumstances. In stage 2 (associative stage) the relationships learned in phase 1 are practiced until they become smooth and accurate. In stage 3 (autonomous stage) relationships are compiled through practice to the point where they can be done "without thinking." Declarative knowledge thus gets transformed into procedural form . . . we lose our access to it and our ability to report it verbally.

In this model, whatever could once be said about what we do gets transformed into what is no longer said. This process is what I have referred to as internalization.

A similar picture can be obtained from a phenomenologically introspective account of learning the skill of jazz piano (Sudnow 1978, pp. 8–10). Thus Sudnow talks of learning to play chords as follows:

Anyone who has witnessed or been a beginning pianist or guitarist learning chord production notices substantial awkwardness. A good deal of searching and looking is required at the outset. The chord must be detected, first seen as a sequence of named notes taken in with a look that reviews the terrain up and down, finding the chord as a serial ordering of these and those particularly identified tones, going from left to right or right to left, consulting the rules to locate places. Then some missing ones in the middle are filled in. . . . As my hands began to form constellations, the scope of my looking correspondingly grasped the chord as a whole, a consistency developed in seeing not its note-for-noteness, but the pattern of its location as a configuration emerging out of the broader visual field of the terrain. . . . Sitting at the piano and moving into the production of a chord, the chord as a whole was prepared for as the hand moved toward the keyboard . . . the musician . . . in the course of play may see past [the keyboard] into the music with a look that is hardly looking at all.

This part of Sudnow's account describes what I have called the internalization of machine-like acts. He is describing how an initial set of explicit rules, which can only be executed by a human with great awkwardness, come to be executed smoothly as they cease to require conscious thought.

Though smooth execution may be a necessary precursor to the act of jazz piano playing, what we see described in the above paragraph does not tell anyone how to play jazz piano; it only tells them how to play like a mechanical device. What is more, there is no reason why a mechanical device programmed with the behavioral coordinates of what Sudnow knew when he first sat down to learn should not reproduce this aspect of playing without loss. That Sudnow gets smoother and more slick as he ceases to think about the application of the rules is a matter of the way Sudnow, the human being, is made. It just happens that when we humans think about what we do we tend to do it clumsily. This is a matter of psychology and physiology, *not philosophy;* the two are sometimes confused.

Practiced use of a slide rule is another example of internalization that carries through the theme of arithmetical skill from the last chapter. It is easy to teach a novice (who is otherwise culturally competent) the mechanical part of multiplication and division with a slide rule using nothing but verbal instructions such as the following:

To multiply two numbers *a* and *b* whose product is between 1 and 10, place the 1 on the *C*-scale opposite *a* on the *D*-scale. Read the answer on the *D*-scale opposite *b* on the *C*-scale. For numbers whose product is larger than 10 but less than 100, repeat as before, but use the 10 on the *C*-scale instead of the 1. For numbers whose products have different orders of magnitude use whichever of the 1 or the 10 gives an answer. Don't worry about it, the wrong one won't work. To divide *a* by *b,* put *b* on the *C*-scale opposite *a* on the *D*-scale. Read the answer on the *D*-scale opposite whichever of the 1 or the 10 on the *C*-scale gives an answer. Sequences of calculations can be done by "storing" intermediate answers using the cursor as a marker. These procedures give you only the string of digits belonging to the answer; you have to decide where the decimal point should go by working out roughly what the answer should be.

Readers who have never used a slide rule might like to try out these rules. They will rediscover most of the features of the arithmetic-using-a-calculator described in the previous chapters. Novices will find that the really hard part is adjusting the slide and reading the scales, not following the rules.

Practiced slide rule users can do long sequences of multiplications and divisions without thinking about what they are doing. As well as mastering the physical and perceptual skills of moving the slide and reading the scales, they choose the 1 or the 10 without noticing, and swap between multiplication and division without conscious thought. Nothing, however, would be lost from that part of slide rule use described by the rules, were the practiced user to be replaced by a mechanical device. What is more, if the rules were encoded in a suitable language, a slide rule expert system would be able to explain to a novice how to do the same manipulations.

Skill and Explicit Rules

In the above accounts of machine-like action nothing essential is lost when a skill is reproduced by referring only to what can be said about it. Most examples of the execution of a skill are much more complicated. Dreyfus (1979) used Wittgenstein's analysis to show that rules of action could only be fully explicated "on pain of regress." Dreyfus and Dreyfus (1986), turning their attention to expert systems, develop a model of skill in which explicit rules have no function at all in the practice of an expertise.

The Dreyfus and Dreyfus model has five stages of skill acquisition. Only the first, "novice," stage involves self-conscious attention to articulatable rules. They call these "context-free" rules, because in applying them the learner does not take into account the nuances implied by different conditions of application. For example, they say that a context-free rule of car driving would be to change gear at a certain speed, whereas a more experienced driver would take into account the density of the traffic or the use of gears in turning a corner. In stages two to four of the Dreyfus and Dreyfus scheme whole scenarios are more and more used as cues (e.g , for when to change gear in a car), and the role of conscious choice and analysis becomes minimal.[6]

By stage five, "experts" do not experience conscious decision making at all. For instance, in car driving, "The expert driver becomes one with his car, and he experiences himself simply as driving, rather than as driving a car.... When things are proceeding normally, experts don't solve problems and don't make decisions; they do what normally works" (pp. 30–31). Dreyfus and Dreyfus say that this last stage of competence is achieved in car driving by recognizing each road scenario as the same as a situation that has been encountered previously and responding in the familiar way.[7]

Dreyfus and Dreyfus associate self-conscious rule following with naivete, and unselfconscious internalization with fully acquired expertise.[8] This is a correct description for many cases of the way humans move from novice to expert. For machine-like actions, however, the process has no philosophical significance and no bearing on the development of expert systems. What is more, there are other skills in which humans continually make use of explicit rules even at the highest levels—the Dreyfus and Dreyfus model is too monolithic. Finally, their model does not fully account even for fully internalized nonbehavior-specific action because it depends entirely on reference *back* to previous occurrences of similar situations.[9] Real expertise is sometimes a matter of seeing a new situation as an old situation, sometimes a matter of responding to new situations in a new way, and sometimes a matter of *creating* the socially legitimate or effective response. This can only be described as a matter of referring back to previous occurrences on pain of tautology. The Dreyfus and Dreyfus model is better thought of as a description of the development of *experience* rather than expertise. Again, the problem comes partly from treating expertise as a property of the individual, rather than interaction of the social collectivity. It is in the collectivity that novel responses become legitimate displays of expertise.[10]

In the next few pages I describe a number of everyday examples of the use of rules and expertise. It is mundane examples of this sort that expert systems are meant to encode. The examples will help to disentangle the themes of interpretation, internalization, novelty, and self-conscious application of rules in different domains. The limitations of expert systems can then be reassessed in the light of the many different ways in which explicit rules relate to skills.

Examples of Rules and Expertise

The *Arizona Daily Star* of Saturday, May 31, 1986, carried the following story.[11]

> CHICAGO—A rookie bus driver, suspended for failing to follow correct emergency procedures when a girl suffered a heart attack on his bus, was following overly strict rules that prohibit drivers from leaving their routes without permission, a union official said yesterday.
>
> "If the blame has to be put anywhere, put it on the rules that those people have to follow" [said the union official].
>
> A transit authority spokesman defended the rules. He said "You give them a little leeway, and where does it end up?"
>
> [The driver] will be suspended for three days, will receive instructions on the rules and procedures during that period and will have to "demonstrate an understanding of those rules to be reinstated."

In this case, the driver had not interpreted the rule about leaving his route in what we count as a correct manner. He had, as one might say, "driven to rule" (see below). The remark of the transit authority spokesman shows, however, that the correct interpretation is not solely a matter of expertise. There is room for negotiation over what counts as correct interpretation. The bus driver's deficiency was not a matter of expertise, but rather a matter of deficient socialization in moral priorities. The driver was well-adjusted to route following in the world of concerted behavior, but not in the world of concerted action.

The transit authority will try to repair the driver's deficient socialization by a program of training and instruction but, like a spy trainer, they will succeed in making good only those problems that can be anticipated and explicated. No doubt, the driver will know what to do next time a passenger suffers a heart attack—the transit authority will be able to "hack around" that problem—but he or she will not be able to achieve competence in the world of concerted action without further socialization. How the driver could demonstrate understanding of the rules is an interesting question because what we mean by understanding is proper response to unanticipated circumstances.

The fact that rules do not carry with them the rules of their application is put to good use in the industrial action known as a "work to rule." Working to rule is adopting a self-consciously pedantic interpretation of rules.[12] For example, a teacher, hoping to control his pupils closely, distributed a rule book including the instruction: "Pupils must walk in the corridor at all times." Crafty pupils took this as an invitation not to attend classes, instead spending all their time walking the corridors exactly as instructed![13]

More seriously, working to rule is used in Britain as an effective form of industrial action that is difficult to sanction because employees seem to be doing exactly what is asked of them. Pedantic interpretation of conditions of service—especially rules concerned with safety—leads rapidly to inefficiency or paralysis. Except under the specially designed circumstances that allow for machine-like action, smooth running of bureaucratic organizations, including factories, depends upon employees interpreting rules with a very unmachine-like degree of flexibility.

Some of these distinctions can be found in sports coaching. For example, Knapp (1963, p. 3) distinguishes between correct "technique" and mastery of a skill: "Technique may then be defined as that pattern of movement which is technically sound for the particular skill and which is an integral part but not the whole part of that skill." Knapp makes clear that too much concentration on technique may detract from the ability of sportsmen and -women. Full realization of the skill requires so much concentration on judgment and timing as not to leave room for thinking about technique. In short, correct technique must be internalized before true ability can be learned. Notice that correct technique is not always redundant once it has been internalized. Rules of technique, which correspond to machine-like actions, may still be in operation in some unselfconscious sense even when the sport is being played at the highest levels. For example, a skilled golfer might well use the same grip as was learned in the very first lesson. Nothing is gained by saying that the rule of gripping a golf club no longer operates when a high stage of skill has been attained even though we do not have to think about it. The novice's grip and the professional's grip can be described by the same behavioral coordinates. On the other hand, correct technique must be supplemented with a much more flexible approach to the game if mastery is to be achieved.

Establishing New Interpretations

But there is even more to expertise than knowing how to interpret rules. There is also the matter of *establishing new interpretations.*[14] Even in highly ritualized, fully consensual, and long regulated sets of activities such as sports, new interpretations that cannot draw on past experience are occasionally required. The following example happened on June 21, 1986, as I watched a soccer match on television.

In the World Cup quarter-final between Brazil and France the game was still drawn after extra time had been played. The game therefore had to be settled by a penalty shoot-out. One of the French players struck the post with his penalty kick and the ball bounced back. In the normal way the ball would have been counted as dead after it hit the post and rebounded away from the goal; a French player would not be allowed to score from such a rebound. The ball, however, hit the Brazilian goalkeeper, who was a yard in front of the goal line, and was deflected back into the goal. There is no rule to cover this circumstance. Though the ball might be deemed to be dead after it hit the post, the ball did enter the goal because of a goalkeeping error. Disallowing the score would seem to go against the spirit of the game. In the event, the referee awarded a goal.[15]

Coaching Rules

In some cases rules are used to start the process of learning but these rules are not internalized; rather they are abandoned and replaced with a new and different set. I call rules that are used to begin with but that are later abandoned, "coaching rules." To go back to mathematics, coaching rules in algebra for solving equations include "change side, change sign" and "cross multiply fractions." With a more mature understanding of algebra these rules can be abandoned for a more all-encompassing rule: "Do to one side of the equation what you do to the other." This allows for unanticipated operations to be carried out without the formulation of a special rule for each.[16]

Knapp (1963, p. 22) also describes circumstances in which coaching rules *are* left behind when a skill is mastered. She writes:

> There are dangers inherent in demonstration. When a person demonstrates he tends to show what he thinks is done instead of what actually is done in the real situation.... Gilbreth [and Gilbreth (1919)] ... found that bricklayers when doing identically the same type of work had three different sets of motions. Set 1 was used to teach the beginner, set 2 was used when working slowly and set 3 when working rapidly. At one time many tennis coaches used to demonstrate the playing of a low ball at tennis with a method never in fact used by top-class players in match play.

Tangential Rules

Another kind of rule that has been overlooked is specific to human psychology and physiology. These are rules that do not describe performance but prescribe methods for preparing the human mind or body to achieve success. Rules about the "inner game," such as visualizing success before execution, are an example. Sometimes the advice looks like an ordinary coaching rule even though it could not be applied literally. For example, in golf there is an adage "let the club do the work." Other rules of the golf swing concern the follow-through of the club, yet advice on the follow-through can apply only to the human organism, because a machine could be built to hit balls perfectly without any follow-through at all; the follow-through is irrelevant because the ball has left the club before it happens. Another useful rule of golf is to hum "The Blue Danube" while playing a round. Such a rule could have no relevance to a golf-playing machine; these rules apply only to the human organism. For the human, concentrating on the follow-through helps the rest of the swing, and humming "The Blue Danube" helps to keep the rhythm slow and smooth. I call these pieces of organism-specific advice "tangential rules."

Tangential rules, designed as they are for the human organism, would not be a good starting point for programming a machine to mimic human action though they would be vital parts of an expert system that was to replace a human expert in the role of teacher. This distinction is easy to see in the case of physical actions, such as the golf swing, but the same applies to mental activities such as arithmetic. For example, it is a good rule for humans to work out a rough answer to a problem before working out the answer in detail so as to avoid major blunders. Such a rule has no application in a machine.

Learning Without Rules

It should not be forgotten that we learn most of the foundations of what we learn, including the ability to speak, without any instruction at all in the form of explicit rules. This, of course, is the heart of the phenomenological critique and it is why the notion of socialization remains so stubbornly irreducible from Semipalatinsk to the TEA-laser.[17] For abilities that are regularly learned without reference to explicit rules, knowledge elicitation does not seem to be a sensible place from which to start building a knowledge base.

Different Concepts of a Rule

At the other extreme to tacit rules there are rules belonging to science that can reproduce certain abilities without corresponding to human activities, conscious or unconscious. Consider this playful comment: "The rugby fullback has to know his coordinate geometry. His approximate locus for the wider conversion must be a rectangular hyperbola on which the kicker must place his ellipsoid of revolution before sending it on its parabolic way."[18] We may be sure that this does not describe the rules that rugby fullbacks follow though the physics is a nice literal interpretation of coordinates of action; the physics could be used to design a ball-kicking machine. As for what the rugby fullback knows, he can say more about how he manages to kick than about how he manages to speak. The following is a quotation from Jonathan Webb (1988), the England fullback, talking about kicking goals.

> My ritual, my routine? Heel out the circle, then the trough. [In rugby a hole is made in the ground to support the ball.] Always adjust the ball so the nozzle is slightly to the right, just so, then angle the upright torpedo just slightly towards the posts for more forward impetus. Eye alignment, Inner calm. Consider the wind. Stand. Left foot alongside ball, right instep right behind. Visualize the kick sailing over. Eye and foot aligned. Wipe hands. Four precise steps back. Stop. Check. Visualize. Then two-step chasse to left if it's a Mitre ball that we use at Bristol, one-and-a-half if it's

a Gilbert at Twick[enham]. You must time them better though they go further. Visualize how it will feel on your foot. . . . If you tried to write down on paper exactly what you do to kick a ball between two posts with absolute certainty, it would be impossible, you'd still be at it in a million years—but once you've done it just once, your body and mind has the exact formula stored and ready to be repeated.

Webb articulates many of the points already mentioned, including tangential rules, the impossibility of writing down in full the skills of goal kicking, but also certain articulated rules that he still uses; note especially the rules about number and pattern of steps and the way this varies according to the type of ball. Reproducing these rules in an expert system, though it would not be a very good way to build a goal-kicking machine, would nevertheless make for a useful teaching machine; this must be so for reasons no more profound than that coaching manuals, full of explicated rules, are of at least some use.

Articulated Rules Used by Experts

The matter of articulated rules is even more obvious in golf. I have already mentioned tangential rules of the golf swing, but explicit rules with more direct application continue to be used to maintain and improve golfers' play up to the very highest levels. Golf professionals are continually analyzing their swings and returning to their coaches for help. For example, in the autumn of 1987, the German professional Bernhard Langer was having difficulty hitting the ball straight. Fellow professional Ian Woosnam advised him to move the ball a couple of inches forward in relation to his stance. Langer reported (to the newspapers) how relieved he was to see the ball flying straight again as a result of that simple adjustment. Woosnam himself was later to suffer a catastrophic loss of form:

His form has been so bad that he has been using every moment of the day to try and find out what is wrong and, late one night, swinging a club in front of the mirror, he decided that not enough of his weight was on the right foot.

"In practice," he said, "it works. I'm taking big divots again and that's how I like to play." Now he has to try it on the course. (*Guardian*, April 6, 1988)

About a month later, the same reporter followed Woosnam in a tournament:

Woosnam faced a tester of his own, a three-footer at the 16th, which he fairly hammered into the hole, and then he produced the stroke of the round, a lovely drowsy six-iron from the middle of the 17th fairway which came down seven feet from the hole.

"It's been a long time since I hit shots like that," he said afterwards, and they are indeed the hallmark of his play when the rhythm and confidence are there. (*Guardian*, May 7, 1988)

Golfers nearly always report their state of form in this passive way. They say "I'm hitting the ball well," or "I'm not getting into the ball," or some such, as though the cause was outside themselves. And this passive mode of expression is appropriate since changes in a golfer's form can often be cured by small technical adjustments made *to* them rather than *by* them, as it were. Of course these rules, however much they are referred to, cannot be the *whole* basis of golfing skill, or a beginner with substantial book learning would be as good a golfer as a professional. Nevertheless, statable rules are a component of this skill.[19]

(In passing we might return to Semipalatinsk and note that the last quotation from the *Guardian* illustrates another feature of the difference between what we can come to know through talk and what we can know through socialization and practice. In the second paragraph the writer refers to a shot as "drowsy." I am sure I had never heard this usage before reading the story in the paper, and I am sure if I were trying to train someone to "pass" as a golfer, I would not have thought of using the term in the context of golf before May 7, 1988. I believe that a nongolfer, who had learned everything about golf from books or conversation, would not be able to infer the meaning of "a drowsy shot" during interrogation. A golfer, on the other hand, will know what it means without being told.)[20]

On the matter of the use of articulated rules, what applies to sport is equally true of other familiar skills. Though learning to speak English has little or nothing to do with articulated rules, learning written English style has something to do with knowing rules such as: "Don't split the infinitive"; "Never end a sentence with a preposition"; "Never put more than one idea in each sentence"; and so forth. Skilled users of English—who learned to speak and write without conscious teaching of grammar—may still improve their stylistic performance by learning rules of style, and trying to keep them at the forefront of consciousness.[21]

Knowledge Elicitation and Expert Systems

The above examples show that the relationship between what we know and what we can *say* about what we know is complicated. The Wittgensteinian point, that no set of rules can ever replace everything we know, does not lead straight to design principles for expert systems. This is because the relationship of people to rules is complex. It is complex because systems fit into society like artificial hearts fit into bodies and the "body" of users does not always need everything to be made explicit; it is complex because the degree of explicitness of rules in expertise varies from skill to skill; and it is complex because the relationship between being able to exercise a skill and the rules used for training vary markedly from area to area and person to person.

Drawing on the above examples we can work out the implications for knowledge elicitation-based expert systems. Knowledge elicitation is widely quoted as being one of the more troublesome aspects of building expert systems. The problem is cited in the introductory paragraphs of many papers. Here is an example from a paper by two of the leading British experts on knowledge elicitation:

> The power of an expert system depends on how much knowledge it has and how good that knowledge is. The process of acquiring this knowledge from the relevant human experts and representing it adequately within the system is thus a crucial phase in the construction of an expert system. Unfortunately, experience shows that "knowledge acquisition" is both a difficult and time-consuming task and is proving to be a major bottleneck in the production of expert systems. (Kidd and Welbank 1984, pp. 71–80)[22]

Many of the supposed practical difficulties of knowledge elicitation are not practical difficulties at all; they are a matter of mistakes in the analysis of what people know about what they know. Assume absolutely optimum conditions for knowledge elicitation. What is the best that could possibly be done in getting useful knowledge from the heads of experts for putting into machines? It depends on how rules were used in developing the skill in question.

Use of Rules in Development of Skills

Rules relate to the practice of skills in the following ways:

1. Application of Formal Rules

1a Rules are learned and consciously used to exercise the skill (within a matrix of filtering and interpretation, of course).
example: novice players of card games; novice user of a slide rule; novice engaged in machine-like actions; inefficiently working to rule.

In these circumstances knowledge elicitation will provide all that is necessary to form the basis of a computer program that will recapture the skill. The elicited rules can be encoded in a mechanical-logical equivalent to the behavioral coordinates of the action.

1b Rules are learned and then become internalized.
example: the constitutive rules of chess (that is, the rules that make chess, rather than the rules of good play); practiced use of a slide rule; the golf grip; practiced machine-like action and reflex-like abilities.

The implication for the knowledge engineer of this relationship is that an expert is not a suitable source of knowledge. A better procedure would be to interrogate a novice who has not yet internalized and "forgotten" the rules. (Or interrogate an expert under circumstances—such as conflict or a specially difficult case—that will force internalized procedures back to the surface.)

In this or the previous case it is reasonable to assume that a machine will use rules in at least as quick and efficient a way as a human. We would expect that machines programmed on these principles would be very good at those aspects of skill that involve internalizing the constitutional rules of games (e.g., the moves of the chess pieces, the rules of card games), very good at using a slide rule (provided the robotic and visual aspects of the problem could be solved), and excellent at working to rule in a factory. Knowledge elicitation could provide the basis of the programs for all these tasks. What is more, machines based on knowledge elicitation would make good teachers of their embodied skill. Recapitulation of the rule base would tell the novice all he or she needed to know. The novice would then be in a position to become an expert through practice and internalization.

2. Coaching Rules

Rules are learned but must be forgotten as skill is acquired so that something different can be internalized.
examples: the "low shot" at tennis; bricklaying; Dreyfus and Dreyfus; aspects of car driving.

Here knowledge elicitation is potentially misleading. A system based on novices' rules might go some way toward mimicking what an expert consultant would say, but could only go as far as the spy who had been taught by purely explicit means. This is because, ignoring the knowledge engineer for the moment, all that can be put into the program is what the expert knows how to say, and this no more captures advanced expertise than what the native of Semipalatinsk can say captures Semipalatinskness.

3. Tangential Rules

Rules that do not describe the action itself, but do help the human organism to execute the action are learned, and perhaps come to be used unselfconsciously.
examples: "let the club do the work"; "work out a rough answer to the problem before working it out in detail."

In this case knowledge elicitation will be a positively misleading beginning to a program because machines are so different from humans. This may explain some of the classic problems of the relationship between knowledge engineers and experts, which lead the engineers to complain that the experts give them misleading advice. For example, in the chapters that follow it will be seen that expert crystal growers would be likely to advise an inquirer of the importance of scrupulous cleanliness and accuracy. These rules are frequently ignored in practice—they have to be—yet they may instill a useful attitude of mind—that is, be clean and accurate unless you can't, which is a good guide to the novice, a much better guide than "be slovenly and haphazard." It makes the novice attend to particular circumstantial reasons for not being clean and accurate in a way that the rule "don't bother about cleanliness and accuracy," would not. "Be scrupulously clean and accurate" can be seen as a tangential rule of crystal growing.

4. Internalization Plus Self-Conscious Maintenance and Adjustment

Examples: Rugby fullbacks; professional golfers.

In this case, if the skill consists of internalized machine-like rules, as in 1b, then knowledge elicitation would make a very good starting point, just as in the case of 1b. If, on the other hand, the internalized rules were either coaching rules or tangential rules, then knowledge elicitation could not form the basis of a program that would reproduce the skills in question, but it could produce an expert advisor programmed with the explicit rules used to adjust and maintain the skill. Actually, such systems are increasingly being built . For example, the latest generation of medical expert systems are much more medical *advisors*, which leave most of the skill in the hands of the doctor while providing reminders, updates, and other help based on more formal data manipulation.[23]

5. Socialization Without Conscious Rule Following

A skill is learned entirely by enculturation, and conscious rules never play a part.

examples: Speech comprehension; pattern recognition (A subset of machine-like actions may be learned this way. They would be mastered directly at the reflex-like stage, without conscious training or conscious restriction of freedom.)

In this case knowledge elicitation can only play a part if the program is no more than a compilation of examples. It will be noted that knowledge elicitation is not used in speech or pattern recognition programs for, though we are all experts in these skills, we are unable to provide any general rules of practice. We can, however, provide an endless string of examples of good and bad practice.

6. Socialization Without Rules Plus Conscious Adjustment

A skill is learned by enculturation but higher levels are attained by the exercise of learned rules.

examples: Written English style; elocution lessons.

Here, just as in the case of "4. Internalization Plus Self-Conscious Maintenance and Adjustment," knowledge elicitation cannot give rise to a program that can reproduce the skill, but it could form the basis of a good mechanical advisor.

It is only after these relationships between the use of rules and the exercise of expertise have been understood that the practical problems of knowledge elicitation can be tackled.

7

Rules and Expert Systems

We know that we can never approach full articulation of our abilities because the words would ramify indefinitely. Going back to the examples of the well-developed constitutive rules of soccer and cricket discussed in the last chapter, we found that even in those cases of very well-established sports, with highly codified constitutive rules, played at the highest standard and with the careful planning that befits world class events, the rules were not adequate in themselves to cope with unforeseen circumstances. In such circumstances we usually think of the rule base as *deficient*; we think that the rule base is potentially perfectible given enough time, experience, and dedication. But a perfect set of rules would have to cover every possible application. This is to say little, for such a set of rules is a complete description of the past and of every possible future.

Yet, although all occasions of use of a rule cannot be foreseen, and although the application of a rule on a particular occasion may depend on humans' rights and abilities to *make* precedents, sets of rules do get better. In the case of the World Cup, a precedent has been set; the next time the ball hits the goalkeeper in a penalty shoot-out the referee will know what to do without having to *intervene* in quite the same way. Although it is not correct to think of the rules of soccer as deficient because they are not perfect, it is possible to improve upon them so as to eliminate at least some occasions of ambiguity. If this is true of highly codified rules, it is even more true that what we can say about a skill is not fixed once and for all—what we can say develops and changes as science progresses, as we become better at introspection, or as external events force us to think harder about what we normally do without self-conscious guidance. The idea of tacit, inexpressible knowledge doesn't leave enough space for the fact that more and more detailed descriptions of actions can be extracted from us. We can hack away at lists of rules within rules, lists of exceptional cases and, in the last resort, we can provide a more and more exhaustive list of examples. To deal with this we need a model more akin to the world as seen by the hacker, if not so beset with false hopes. For this, I return to that *reductio ad absurdum*, the indefinitely large set of rules and examples.

The "Rules Model" of Culture

In what I call the "Rules Model" of culture, human beings are programmed with rules that would tell them what to do in every conceivable, if not yet conceived, circumstance, and rules within rules to tell them what to look for to recognize those circumstances, and rules within those rules to tell them how to recognize the things that help them recognize the circumstances, and so on *ad infinitum*. Obviously, humans don't "know" these rules but we will think of them as somehow being able to draw on this resource. This, for the sake of argument, is what I shall take their cultural, interpretative, and inductive competences to be. We can think of the formal and statable content of human knowledge as resting on a foundation of perceptual and cultural interpretative abilities, but we will think of this foundation as itself made of endlessly ramifying rules that remain unexpressed.

The Rules Model, in case I have not made it clear, is not a realistic model.[1] I may be able to teach a rule of arithmetic as a formal rule, but I could never teach *all* the rules of arithmetic as formal rules. For me to teach a rule of arithmetic, the recipient already has to know a great deal of arithmetic—that is, the recipient must know how to use numbers, their invariance with respect to exact shape, the time of day, the day of the week, the weather, the state of one's health, and so forth.[2] We encountered the spirit of the Rules Model in discussing automated paint spraying. The act of paint spraying, if it were broken down into separate machine-like acts, could only be reproduced without loss by an indefinitely ramifying number of such acts so that, in effect, the whole future of chair spraying were anticipated.

Real or not, there are at least two things that make the Rules Model useful. First, the model fits well with the way some knowledge engineers think about human knowledge. For example, Hayes-Roth (1985) suggests that "A commercially practical system may require as few as 50 rules," that "expertise in a profession requires about 10,000 rules," and that "the limits of human expertise are at about 100,000 rules."[3]

The second useful feature of the Rules Model is that it explains why people expect that as computers get bigger and better, and contain more and more rule-like instructions, they will come to imitate humans more closely. This vision of incrementally increasing approximation to human abilities is familiar enough and has some validity so long as one is on the right side of the "combinatorial explosion."[4] Given clever and increasingly complex programming, in ever larger and more powerful machines, we can hold off the moment when the difference between action and its emulation via behavioral coordinates thrusts itself upon us. Although Winch points out that no coordinate description, however detailed, can substitute for the statement about a cat writhing, and although Searle is at pains to point out the in-principle differences between action and behavior, we can see how a charitable observer might mistake one for the other.

Ramifying Rules and Expert Systems

One of the terms used in the expert systems literature is "heuristics." This literature reveals a burgeoning range of knowledge typologies (for example, see Welbank [1983]) but their almost universal characteristic is a stress on the dividing line between factual information and experts' informal rules of thumb, or heuristics. Because heuristics are rarely written down, and because the knowledge engineer has to extract them from the expert in a painstaking manner, they are seen as crucially different from readily codified facts. Heuristics are sometimes taken to exhaust the informal aspects of what is known. Heuristics, however, are closely related to facts and formal rules because, like them, they can be articulated. The crucial dividing line is that between explicable rules and facts, and the nonexplicated component of knowledge upon which even formal rules and facts rest.

We can use as simple a program as PICKUP (chapter 6) to see what happens to an expert system if we try to explicate the foundation of the rule base. Imagine what would be required if PICKUP were to be expanded so as to be usable by someone without normal Western cultural competences. Two of the questions PICKUP asks concern "fleeting smiles" and "eye contact." Suppose the meaning of these was not clear to the user. The explanation required could go roughly as follows (I ignore PROLOG's syntax for this purpose, but all these rules could be easily encoded if necessary):

A smile is fleeting if it lasts for a short time.

A short time means less than one second, but long enough to be clearly recognizable.

A smile is a movement of the facial muscles so as to raise the corners of the mouth and the eyes.

A smile is not a tic.

A tic is an involuntary movement.

A voluntary movement can be recognized because it is not repeated regularly and is not made in a jerky manner.

A jerky manner implies a sudden start and stop to the action.

Sudden means in less than about one-tenth of a second [it is important for a successful pickup that these time intervals be judged rather than timed—e.g., by stopwatch].

Eye contact means contact between pickup target's eyes and user's eyes.

Eye contact does not mean physical touching of eyeballs.

Eye contact means that the pupils of the eyes should be aligned in such a way that beams of light passing normally through the user's pupils would converge on one of the target's pupils and vice versa [this should not be tested by experiment].

The pupil upon which the beams would converge should alternate between left and right but not regularly

Proximity between target and user increases likelihood that eye contact is being achieved except when contact is made across a crowded room.[5]

In a crowded room the weighting of eye contact as signifying success is increased.

Prolonged means . . .

Etc.

Though extending the rule base in this way makes the program in principle more accessible to a wider set of potential users, the developer of expert systems will experience ramification of the rules as a sort of technological nightmare. If we add an extra clause to PICKUP to cover just part of the definition of "fleeting," then the new program looks as figure 7.1. A new consultation is shown below the figure.

```
X will-come-back-to-my-place if
        X has-a-fleeting-smile and
        X will-maintain-prolonged-eye-contact and
        not X is-short-sighted

X has-a-fleeting-smile if
        X smiles-for-less-than-one-second and
        not (X starts-smile-in-less-than-one-tenth-second) and
        not (X stops-smile-in-less-than-one-tenth-second)
```

Figure 7.1
Extended pickup for Martians

```
find (X:X will-come-back-to-my-place)
WHICH (X : X SMILES-FOR-LESS-THAN-ONE-SECOND) ?
ANSWER IS . Jane
ANSWER IS . Alice
ANSWER IS . Lucy
ANSWER IS . Suzy
ANSWER IS . Carol
ANSWER IS . end
IS IT TRUE THAT JANE STARTS-SMILE-IN-LESS-THAN-ONE-TENTH SECOND ? yes
IS IT TRUE THAT ALICE STARTS-SMILE-IN-LESS-THAN ONE-TENTH-SECOND ? no
IS IT TRUE THAT ALICE STOPS-SMILE-IN-LESS-THAN-ONE-TENTH-SECOND ? yes
IS IT TRUE THAT LUCY STARTS-SMILE-IN-LESS-THAN-ONE-TENTH-SECOND ? no
IS IT TRUE THAT LUCY STOPS-SMILE-IN-LESS-THAN-ONE-TENTH-SECOND ? no
IS IT TRUE THAT LUCY WILL-MAINTAIN-PROLONGED-EYE-CONTACT ?   no
IS IT TRUE THAT SUZY STARTS-SMILE-IN-LESS-THAN-ONE-TENTH-SECOND ? no
IS IT TRUE THAT SUZY STOPS-SMILE-IN-LESS-THAN-ONE-TENTH-SECOND ? no
IS IT TRUE THAT SUZY WILL-MAINTAIN-PROLONGED-EYE-CONTACT ? yes
IS IT TRUE THAT SUZY IS-SHORT-SIGHTED ? no
==> SUZY .
```

It is as though an expert system is a vessel into which the expert pours knowledge and from which the end user takes knowledge. If we think of this knowledge as chicken soup with dumplings, then the expert system is like a sieve; with all known expert systems, the dumplings get transferred but the soup is lost. The dumplings are the readily statable facets of knowledge such as factual information and heuristics that can be made explicit, whereas the soup is the great unarticulatable cultural foundation on which the facts and rules rest. To capture the soup, the mesh of the sieve must be infinitely fine.[6]

Many expert system designers are worried about the size of the holes in the sieve. One might describe a lot of current research as being directed toward making the holes ever smaller so that eventually the whole broth can be transferred without loss. But given the previous analyses, completely closing the holes is impossible. There is a solution, however. If we return to the problems of PICKUP, we notice that the rule base does not explode if the end user *is* culturally competent. In real life, the user of the information will supply much of the interpretation that a finite system of rules cannot supply. The following three case studies show how users supply interpretation.

In a classic experiment Harold Garfinkel (1967) provided students at the University of California, Los Angeles, with an experimental counseling service. The service was free but questions had to be asked of the counselor so that they could be answered with a yes or a no. The students presented written questions and received replies without actually seeing the counselor. Many of the students felt the new service provided adequate counseling. The counselor could have been a rudimentary expert system with a poor output repertoire but, as a matter of fact, the counselor comprised a list of random numbers keyed to the yes or no answers (see also Oldman and Drucker 1985; Suchman 1987). The students in this experiment provided so much interpretation to the "data" that they were able to read sense into noise.

This is a paradigmatic example of the capacity of human beings to make sense of anything including information that isn't there. In the light of Garfinkel's experiment the success of ELIZA is hardly surprising nor is the interactive sufficiency of other simple computer programs. Humans are so accustomed to constructing what counts as sense out of the rudimentary signals that comprise conversation with other humans in a familiar cultural context that they are not easily put off by nonsense.[7]

In the example of the counselor we saw the end user inserting not only the interpretation but also making up some imaginary information to fit it. Suppose the counselor is thought of as an expert system. Then, if one pictures the expert system situated between expert and end user thus:

EXPERT —— EXPERT SYSTEM —— END USER

one might say that all the knowledge was inserted at the right-hand end. This, of course is rather an extreme model of knowledge *transfer* and exaggerates the role that the end user plays.

The opposite case is represented in another study. In the early 1970s I found that scientists who wanted to build Transversely

Excited Atmospheric pressure lasers (TEA-lasers) did not succeed if they used only written sources of information even when the authors tried their best to make certain that the documents contained every relevant fact and heuristic (Collins 1974, 1985). What is more, these scientists were unable to build the laser even after they had engaged in prolonged conversation with middlemen who knew a great deal about the devices but had not yet built one themselves. Even where a scientist had substantial contact with a successful laser builder this would not guarantee success; such contact was a necessary but not a sufficient condition for knowledge transfer. One might say of this case that it was impossible to find a sieve with small enough holes to transfer the knowledge. The end user, if he was to succeed in building a laser from a set of instructions, had first to become a full-blown expert himself. This he could only do by serving something close to an apprenticeship with an existing expert. Those who did not possess the skills themselves could not act as satisfactory masters to apprentices.

In this example of knowledge transfer, *all* the knowledge was inserted in at the left-hand end, and because we have no sieves without holes, it could not be transferred by any kind of intermediate vessel, including unpracticed human beings.

The model of knowledge transfer that arises out of TEA-laser study is again an extreme. The skill in question was very new and end users had not yet learned to interpret the information properly. Fortunately, the areas of expertise that most knowledge engineers try to model are not like this. They are much more familiar so they can rely on the end user providing far more.

A third, in-between, more typical example of knowledge transfer is widely known. It is the use of a recipe in cooking. It typifies the successful use of a written list of instructions to do a task that we could otherwise not accomplish.

I have a recipe for port wine soufflé. Because I have no face-to-face communication with the author of the cookbook I cannot learn skills from her after the manner of an apprentice. Nevertheless, at the end of the day, the recipe will enable me to make the soufflé. This is because of what I, the end user, bring to the interaction. For example, the recipe contains the following instruction: "Beat the egg whites until stiff and then fold in." To manage this I must already know an indefinitely long list of things of the sort we saw was needed to make sense of PICKUP. Thus I must contribute my knowledge of what all the words mean and this includes knowing that a *white* is not white but transparent—at least until it is beaten. I must know how to get the white from the egg by cracking it and separating it from the yolk and the other bits. Some of these other bits, such as membrane inside the shell and the shell itself, may be more white than the white. I must then know how to beat the egg and with what to beat it.[8] If I am not an experienced cook I probably won't succeed in making the whites stiff because one may beat for an unexpectedly long time without significant alteration to the consistency of the white. The inexperienced are most likely to assume they are doing something wrong long before the egg starts to stiffen. Certainly it is hard to imagine that the eggs will ever be so changed in character that the bowl could be inverted without their falling out—but that is what stiff means. Then I have to know the special meaning of "fold" in the context of cooking and so on. In short, I will have to be a fairly accomplished cook at the outset to be able to make use of this instruction.

How did I become so accomplished? The answer is "through apprenticeship." Most of the knowledge I use to cook a soufflé was not transferred via the written recipe but via face-to-face contact in my mother's kitchen.

In the first model of knowledge transfer the expert transferred nothing to the end user, so no intermediate vessel was involved (even though the end user did not realize that this was the case). In the second model of knowledge transfer the expert and the end user shared little or nothing, and that is why no medium of knowledge transfer could work (the only medium that could work would be the elusive sieve without holes). Fortunately, it is the third model that represents the typical situation for the knowledge engineer. Once the nature of this typical relationship is grasped, the relative difficulty involved in building expert systems to fulfill various tasks is easily understood.

Stages in the Development of Expert Systems

Class I Expert Systems

We can apply the Rules Model to generate a simple typology of expert system. From the perspective of the Rules Model the simplest type of expert system to build is one in which the program is based on a set of rules that are already encoded for use. For example, the rules might be encoded in a handbook. This approach guarantees that the necessary foundation of cultural competence is already in place or the handbook would be useless.

Such a design philosophy is applied by, for example, Expertech, a knowledge-engineering firm. They have reproduced the British Government Statutory Sick Pay regulations in an expert system for use by employers.[9] A spokesman for Expertech explained that their aim was to avoid "going over the wear" in the design of expert systems. Going over the wear resulted from tackling problems that were too difficult and spending hundreds of thousands, or even millions, of pounds, without any guarantee of a successful product at the end. My respondent described the Statutory Sick Pay system in the following terms:

> That [showing me a 59-page booklet] currently is the DHSS [Department of Health and Social Security] booklet on Statutory Sick Pay. As you'll see, it's a nice, readable little booklet. . . . Our knowledge base is not intended to be totally definitive, but there's an awful watershed in some situations where you try to make a knowledge base absolutely definitive in every respect. [Compare this with] something which is pretty helpful in most situations. . . . [we're going for the latter] . . . hopefully building checks in, so that it recognizes, or can explain its limits and bounds.

The strategy just described is represented diagrammatically in figure 7.2. The figure leads naturally toward a classification scheme for expert systems. In this scheme *Class I* systems encapsulate only ready-coded knowledge. They refer the user to a *domain expert* whenever their knowledge base fails to match the user's cultural competence.[10]

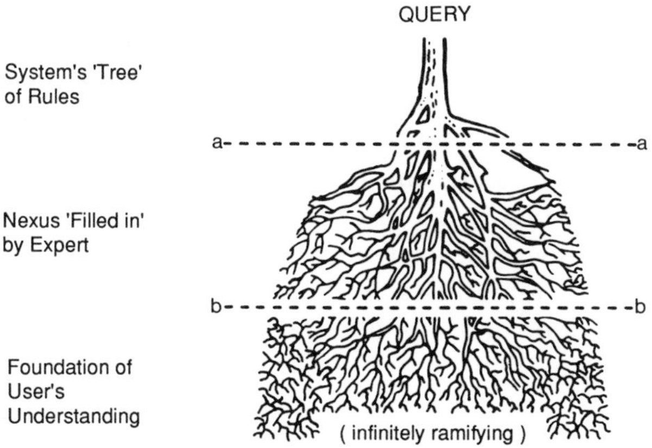

Figure 7.2
The expert fills the gap between machine and user

Class II Expert Systems

A second class of expert system contains a rule base built up largely of rules of thumb or heuristics. These are rules of expertise elicited from domain experts by detailed questioning. The early and widely publicized expert systems used such rules. The famous system MYCIN[11] used medical diagnostic rules, and another celebrated system, PROSPECTOR, was designed to use rules for the interpretation of seismic signals to locate deposits of valuable minerals. There is no doubt that these *Class II* systems are much harder and more expensive to make than Class I systems. For example, Karl Wiig, of the American financial consultants Arthur D. Little, has suggested that developing a basic system for use in the financial planning environment would take 10 to 20 person-years while deploying a full system might take 30 to 150 person-years. Obviously the costs involved are to be measured in millions of dollars.[12]

In spite of the cost and difficulty, in terms of the Rules Model the problem is not different in kind from the problem of building systems from ready-encoded rules. This is because, so long as the knowledge engineer sticks to rules belonging to the esoteric knowledge base of the domain expert, and so long as these can be articulated, and so long as lay end users will consult the machine in collaboration with a more expert advisor (so that tacit knowledge is not an issue), there is no reason why the rule base should explode catastrophically.

Class III Expert Systems

In both Class I and Class II the dumplings are provided in the expert system but the soup is provided by a combination of the end user and an intervening domain expert. *Class III* expert systems are intended to dispense with the domain expert. They encounter problems of a different order of magnitude. The knowledge base of this class may be founded on the ready-coded rules of Class I or the esoteric heuristics of Class II, but it must contain much more because the system has to be used directly by a lay end user. Of course, even a lay user is an extraordinarily accomplished expert, stuffed full of the knowledge soup acquired during socialization; without these abilities none of us would be able to understand any advice however mundane. Thus, provided the system's knowledge base can reach out and touch the user's knowledge base directly (in figure 7.2, the line a——a comes down to touch b——b), then the system ought to work.[13]

We can imagine systems of this sort that deal with knowledge already within the compass of the general public—for instance, we can imagine an effective automated railway timetable and information service—but it would be hard to distinguish such systems from Class I. Class III systems are meant to deliver esoteric skills to the uninitiated.[14]

An example of a system that is intended to make direct contact with the public on matters esoteric is that part of the Department of Health and Social Security (DHSS) "demonstrator" built at the University of Surrey to help people claim social security benefits. The team leader described this project as being aimed at:

> the general public, the people who have to deal with this huge and complex organization and its regulations.... [We] are building systems to demonstrate how present day computer technology could make claiming benefits easier, and the regulations more comprehensible. (Gilbert 1985, p. 2)

One part of the project is a "Forms Helper," intended to aid claimants in filling out forms. The project *is* based on the ready-coded knowledge of the DHSS rule books but, as Gilbert (1985, p. 2) put it:

> [A person filling in a form] has no context to help decide on matters such as how precise the answer should be, whether an estimate would be acceptable, or whether the claim would be invalid if no answer at all is given to a question. These are all areas where the Forms Helper can assist, with examples of possible answers, with explanations of terms, and with syntactical or even semantic checks on the answers for clarity and consistency.

As the quotation makes clear, the Surrey project, like any other project that aims to deal directly with the lay user in an esoteric area, is trying to cope with the absence of rules for the application of rules. In trying to do away with expert interpreters, Class III systems are likely to go over the wear. If the project maintains the goal of making the DHSS rules fully comprehensible to the lay person it will run into the same problems as PICKUP for Martians (figure 7.1). For example, a standard problem for DHSS claims is the interpretation of the term "cohabitation," upon which substantial benefit entitlement can depend. If a woman has an intimate friend who works on a North Sea oil rig for most of the year and therefore visits her only occasionally, is she cohabiting? The program will have to anticipate every such eventuality if it is to avoid the need for expert interpreters or expert creators of precedent.[15]

More Rules Mean More Users

Returning to figure 7.2, for an expert system to work, a——a has to make contact with b——b. Assuming they are initially far apart, there are three ways in which this can happen: b——b can move up as the population of users becomes more knowledgeable—it is insufficient to consider the machine as a free-standing brain because one culture's information is another culture's expertise; the gap can be filled by a domain expert; and finally, a——a can move down. The line a——a moves down as more rules are encoded into the system, as more of the future is anticipated. Thus, the DHSS advisor, given that the cohabitation problem has been anticipated, could now deal with someone whose partner lived on a North Sea oil rig. Likewise, since June 1986, a soccer World Cup advisor would be able to handle penalty shoot-outs under circumstances where the ball rebounds and hits the goalkeeper. A trained spy, though no complete substitute for a socialized native under vigilant interrogation, is nevertheless much better at pretending to be Semipalatinskian than you or I. We know, of course, that extensions of this sort can never anticipate everything.

Let us summarize what has been said with the aid of a diagram (figure 7.3). In this diagram, the number of encoded rules in the system increases from left to right; this is equivalent to the line a——a moving downward in figure 7.2. The line A——A separates systems that contain ready-encoded rules from systems that need to overcome the "great bottleneck" of knowledge elicitation in order to encode heuristics. Most knowledge engineers would expect to

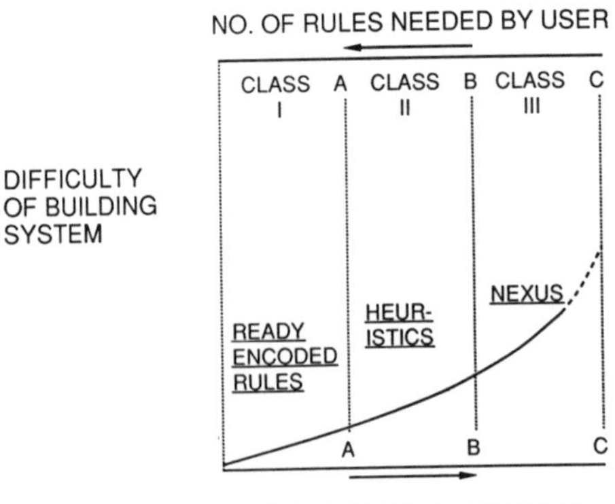

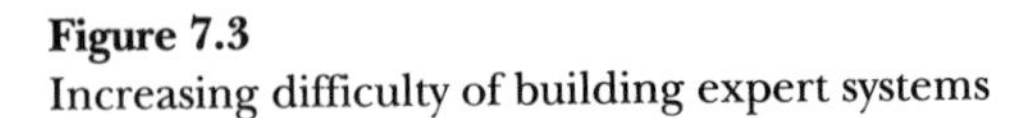
Figure 7.3
Increasing difficulty of building expert systems

see a sharp upward turn in the difficulty of building a system at line A——A, but our argument suggests this is not a particularly significant barrier. The line B——B is of greater significance.

To the right of B——B we attempt to do away with an intervening domain expert and, if the general level of expertise in the population does not increase, this means that we must encode esoteric expertise; at the line C——C the domain expert has been eliminated entirely. This, I have argued, cannot be done.

In these terms my pocket calculator is a usable "expert system," because I can do all the approximating and problem solving that is beyond its competence. I can do this only because I am highly trained in arithmetic. These competences are so widespread that they are not thought of as an esoteric skill.

8

Cultural Competence and Scientific Knowledge

The relationship between what we can say about what we know and what we know is not a static one. What we can say about what we know *increases* as more scientific research is done, as more coaching manuals are written, and even as more philosophical studies are completed. What we can say *decreases* with the incorporation of what was once strange and new into inexpressible common sense—a process that happens at the social level as well as at the level of individual internalization of skill. I now want to analyze these changes.

To do this I will divide up our knowledge and abilities in a commonsensical way into four main categories: facts and formal rules, heuristics, manual and perceptual skills, and cultural skills. As in the Rules Model, we may start by thinking of the first two categories as the expressible part of our knowledge, whereas the third and fourth categories are the unexplicated abilities. We can now look at the way knowledge shifts between these categories.

What were once explicit rules can become part of a society's unexpressed *taken-for-granted-reality* (Schutz 1964), whereas some taken-for-granted ways of going on can be analyzed, expressed for the first time, or reexpressed. Shifts of this sort can be seen by looking at the changing instructions on simple machines in the public domain.[1] For example, an elementary pinball machine, built in the 1830s and to be found in a backroom of the Smithsonian Institution, has instructions that include the following:

'5c A game of skill. Skill hole must be made.

How to operate

1. Place coin or free play token in coin slide and push slide all the way in until balls have cleared then pull slide all the way out.

2. Push RED knob to elevate ball to playing surface.

3. Pull back BLACK knob on plunger and release.

Nowadays, everyone knows how to put money in a pinball machine and how to make the balls run. The 1980s version has only the following rudimentary instructions in the place of what went before:

Insert coin to start machine
Insert coins for additional players[2]

Changes in public understanding may be quite rapid. An 1898 slot machine that bears symbols of fruits displays the instructions:

INSERT NICKEL
PUSH DOWN HANDLE

By 1899, the same model was bare of helpful advice. By then everyone who was going to gamble must have learned how to do it.[3]

In the same way, a postage stamp machine of 1941 includes the now-redundant instruction "Tear off Stamps" beneath the slots from which the stamps appear.[4]

There was a time—in the 1880s—when all coin-operated machines must have seemed strange and sophisticated, when users must have needed assistance to purchase a stamp, or to weigh themselves. I can remember when assistants were present at British launderettes (laundromats) just to show the users how to work the machines. And, just as the Washington Metro first defeated me, the first slot machines must have defeated some of those wishing to part with their money. Now our knowledge is such that not only do we not need written instructions, we might be hard pressed to remember what needs to be said to explain the use of such machines to a stranger. (Would one explain that the stamps must be torn off?)[5]

In spite of this now being part of our tacit knowledge it could, in principle, be reexpressed without loss just as it was in the early machines. Nothing of great philosophical significance happens as the skill required for using these machines switches from articulated form to taken-for-granted form and back again—nothing of greater significance than happens as rules for machine-like actions such as military drill become internalized and vice versa. With rearticulation of the rules these machines would become usable by more distant cultural communities; presumably such machines, when they are installed in countries that do not share modern Western culture, are provided with a greater range of instructions for use. The analogy with expert systems would be downward movement of the line a——a in figure 7.2.

Machines like this are a special case because to interact with them requires behavior-specific action. In general, the third and fourth categories of knowledgeability are less easy to articulate.

Four Kinds of Knowledge

Facts and formal rules include facts of the sort that are readily explained such as "water boils at 100°C," and "in cricket, when the ball is hit across the boundary without touching the ground, six runs are scored." Later on I will discuss the knowledge of crystal growers and of TEA-laser builders where facts and formal rules such as the melting points or atomic weight of specific chemicals, Ohm's law, and some expressible rules of mathematics are found.

Heuristics are explicable rules of thumb and standard practices. In the scientific literature heuristics are not normally written down in the way that facts and formal rules are. There is, however, nothing in principle that prevents me writing down the heuristics that describe the explicable part of my practices. For example, I use the heuristic "change side, change sign" for solving equations. Heuristics are found in sports coaching manuals and in doctoral dissertations. Examples of heuristics from scientific and technological practice include: "In crystal growing always start the melt cooling from well above the putative melting point"; "The tolerances of TEA-laser electrodes are sufficiently large to make it unlikely that the exact shape of the electrodes is the cause of failure."

In crystal growing *manual and perceptual skills* include the ability to recognize the presence of shapes that suggest the existence of crystals within an otherwise undifferentiated lump of solid material. In TEA-laser building, an example would be the ability to distinguish the different sounds that can be made within a prototype laser cavity. Because recognizing things is a theory-laden process, it is not easy to distinguish these abilities from cultural skills even at the analytic level.

Polanyi's (1958) idea of tacit knowledge provides a link between manual skills and *cultural skills.* Polanyi's well-known example of tacit knowledge is the skill associated with bicycle riding. The formal dynamics of balance on a bicycle do not comprise the rules of riding. A rider may know nothing of centers of gravity and gyroscopic forces yet still ride whereas the most expert bicycle engineer may not be able to do so. The rider knows how to ride but without being able to say how. As I have argued, the same tacit qualities apply equally to nonphysical cultural skills such as recognizing significant objects in a confused environment.

Cultural skills include the abilities required to understand and use facts, rules, and heuristics. By this I mean the ability to read and comprehend information and advice, but not only this. Cultural skill includes that which enables, say, an English or American person but *not*, say, an English-speaking Chinese native, to continue the sequence "2, 4, 6, 8" appropriately as "10, 12, 14, 16" on one occasion and as "Who do we appreciate" on a different occasion. The difference between the machine-like performance of central features of arithmetic, and being able to *do* arithmetic is the addition of cultural skills.

Cultural skills include the ability to make inductions in the same way as others in the world of concerted *action*. It is our cultural skills that enable us to *make* the world of concerted *behavior*. We do this by agreeing that a certain object is, say, a Rembrandt or a certain symbol is an *s*. That is how we digitize the world. It is our common culture that makes it possible to come to these agreements, and it is our means of making these agreements that comprises our culture.

Figure 8.1 represents the conventional hierarchy of the different types of knowledge upon which the idea of Western education seems to be based. The figure also can be read as showing the way

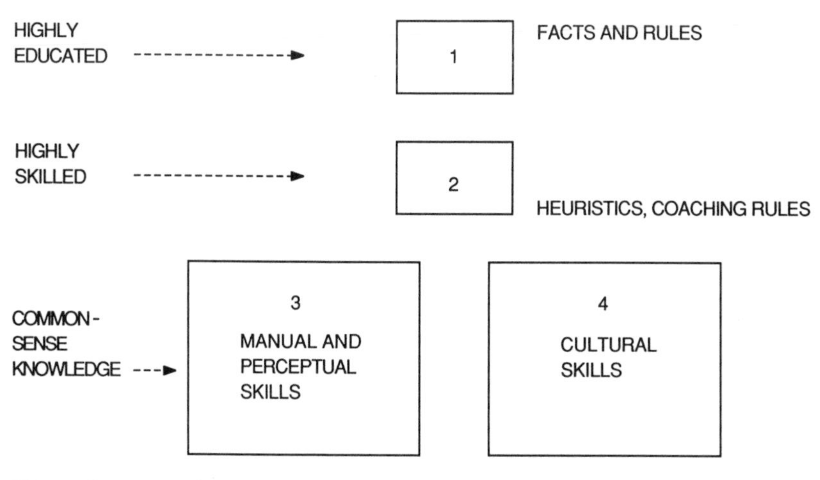

Figure 8.1
The conventional hierarchy of knowledge

that the more formal, explicable categories rest on a foundation of unarticulated skills.

Educational *practices* actually take account of the cultural aspects of knowledge. Thus scientists and mathematicians learn their trade by working through lots of examples under supervision and through guided laboratory experience. Good practice takes into account the dimension of learning that has to do with becoming a member of a new culture—the enculturational model. On the other hand, our *idea* of how we learn subjects such as mathematics is one of transmission of discrete bits of knowledge or sets of self-contained, recipe-like instructions. This is the algorithmical model. The extra bits of practice seem to be there just to remedy defects in the learner. A perfect pupil would read once through the book and then be capable of using all the knowledge therein without loss. One of the ironic implications of the development of computers is that the things we once thought of as the preserve of very clever people have turned out to be relatively easy to mechanize whereas everything else is very hard. This lesson is still being painfully reacquired at the frontiers of new specialisms such as expert systems. Most of us still work as though the most valuable knowledge is that which can be formally, even mathematically, delineated. The whole moment of advance in the sciences seems to be a movement from bottom to top of the diagram (figure 8.1). This, of course, is only a special part of the story.

Examples of Knowledge and Its Transformations

Upward Movements

How does knowledge move between the boxes represented in figure 8.1? One such movement seems to take place when the knowledge of craftsmen is made explicit by scientific research. That is, science seems to shift aspects of knowledge initially in boxes 3 and 4 into boxes 2 and especially 1. For example, the science of metallurgy seems to replace the craft of the ironmaster. The optimism of those trying to build intelligent machines is related to progress in science—the increasing articulation of the world. The more that knowledge moves into the upper boxes, the more can be encoded in explicit form.

As science covers new ground knowledge appears to move out of the craftsman's *private* domain into a *public* arena, independent of local contingencies or esoteric abilities.[6] What was once the prerogative of those who partake in a local world of concerted action becomes part of the much wider world of concerted behavior. Anyone, or any machine, who has learned the steps of the behavioral dance can then join in. In Western societies the behavioral training is ubiquitous, and it is not subject to discontinuities at national boundaries. It is this that makes scientific knowledge appear to be devoid of cultural foundations. Ubiquity looks like universality. Scientific redescription of the world does not remove the cultural base of knowledge, it moves it to a domain in which the cultural competences needed to interpret the knowledge are differently distributed.

To illustrate, consider an example to which I will return later—knowledge of the strength of glass vessels. This knowledge was once available only to the craftsperson.[7] It was the property of local groups who had shared a suitable apprenticeship. Nowadays, however, knowledge of the strength of glass vessels is the property of anyone who can read a chart—see figure 8.2.

I have never served a glass-blower's apprenticeship nor shared a specialist's knowledge of glass, yet by referring to the chart shown in the figure I was able to encode strengths of vessels into an expert system designed to advise on growing semiconductors. Once encoded in the system, this knowledge was able to serve additional equally ignorant users. At some time knowledge of the strength of vessels has taken an upward journey, enabling it to make cultural contact with a wider group of users.[8]

In the mid-1970s I was lucky enough to watch an upward transformation of this sort as it happened. It involved the development of the TEA-laser referred to above (Collins 1974; Collins and Harrison 1975; Collins 1985). I worked with Bob Harrison as he set about building his first TEA-laser. Among the things he knew was that the leads from the top capacitors to the top electrode had to be "short" (see figure 8.3). It turns out that he did not interpret this knowledge properly, and consequently his laser had top leads that were *short* as he saw it but still too long for success. The laser would not work. Eventually, he found that in order to make the leads short enough to count as *short* in "TEA-laser society," he had to go to a great deal of trouble. The large and heavy top capacitors had to be mounted above the bench in an inverted position in a strong and complex steel frame so that their terminals could be very near to the top electrode.

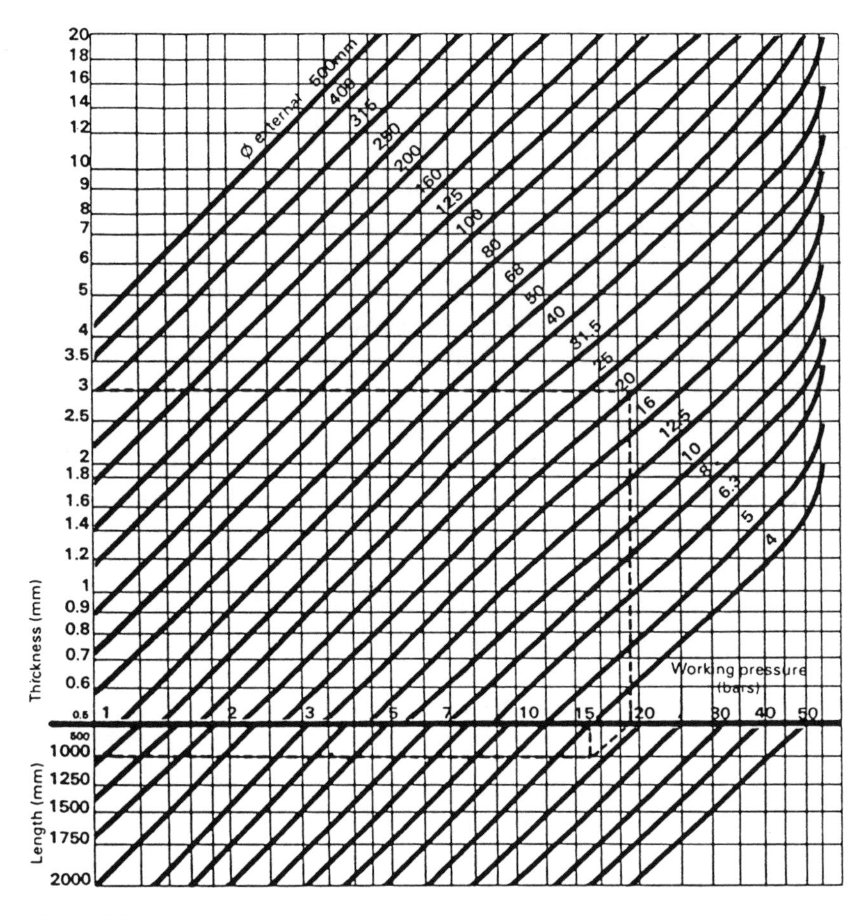

Figure 8.2
Chart showing strength of glass tubes

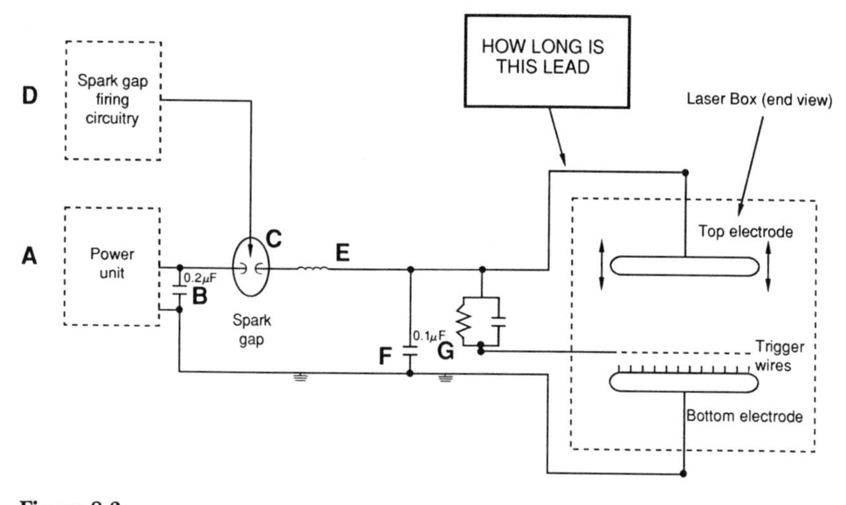

Figure 8.3
TEA-laser circuit diagram

Later, as this aspect of the design of the laser came to be understood as a piece of *electronics*, it became quantitatively theorized. The *inductance* of these top leads was then seen as the crucial variable. It then became accepted that these top leads had to be 8 inches (20 centimeters) or less in length.

Knowledge about lead length can be seen as actually or potentially belonging to three different categories at different times and places. In the first instance Harrison knew the verbal formulation "keep the top leads short" but did not have the cultural knowledge to interpret the meaning of short in the narrow context of laser building. Later, he acquired knowledge about the meaning of "short"; he already knew that *short* did not mean less than one mile (note that we all know enough about the culture of lasers and electricity to see the funny side of that) and he found that it did not mean less than one yard—it meant substantially less than one foot. Within different contexts all these are possible interpretations of *short*, but only the last one is correct in laser society. At the point where Harrison had learned the meaning of *short* where TEA-lasers are concerned, we can describe his knowledge of the length of the top leads as belonging to the *heuristic* category—box 2.

We can also imagine a prior stage (which, as a matter of fact, Harrison did not pass through) in which apprentice laser builders, in copying their masters' design, invert the top capacitors in a strong iron frame as a matter of traditional design aesthetics and thereby have leads of the appropriate length as a matter of course. Such apprentices need know nothing of lead length—it need not even be a concept within their laser building world—nevertheless, there is a sense in which those who could reproduce their masters' skills satisfactorily could be described as having sufficient knowledge of lead length to build a laser. One would have to say that it was part of their tacit knowledge of laser building. Note, that what they knew of lead length would be completely impossible to explicate—firmly in box 4 (with aspects of box 3). Apprentices whose lasers failed to work (perhaps because their leads were too long) would simply be counted as incompetent though no one would know why. This is how correct lead lengths would be preserved without anyone thinking about it. Later, if a rule became articulated such as: "invert your capacitors so as to make sure that your top lead is short enough," then the knowledge would have moved to box 2; it would have become a heuristic.

At the still later stage of history of the TEA-laser, when the matter of lead length was passed into the vocabulary and culture of electronics, the knowledge became formalized—part of the behavioral dance of the numbers. At this final stage, when it was known that the leads must be less than 8 inches in length, the knowledge belonged to the *facts and rules* category, moving from box 2 to box 1.

These shifts of category are complex. They cannot be understood without thinking about the context in which the knowledge is used. Thus, Harrison's first learning to interpret *short* in the appropriate way involved his enculturation; he had to internalize the laser builder's context for the interpretation of *short.* The shift from heuristic to formal rule also involved a shift of context; the top lead moved from being part of the laser builder's art to the world of electronics and arithmetic.

When we reach the point at which the rule becomes "make the top leads less than 8 inches," it appears that we have at last found a decontextualized rule—something universally applicable that is independent of the cultural foundations of knowledge. There is certainly a component of machine-like action in the 8 inch rule, but just like the machine-like actions that can be encoded in a pocket calculator, the apparent universality of this rule depends on the wide (and therefore virtually invisible) distribution of the cultural skills surrounding the machine-like response to the number 8.[9] The number has to be recognized and understood as a number, and it has to be used as a number. Finally, it can only engender the appropriate machine-like response in those who have been trained to respond appropriately. The point is made clear by considering what use the 8 inch rule would be in a more "primitive society." In such a society the notion of "8 inches," or any other length expressed in numerical form, would not appear to have a factual quality at all—it would not be transmittable to others. It would be the private property of those who possessed the esoteric craft of measurement. In such a society, the notion of "8 inches" would be much closer to the bottom of the diagram. What appears to be the same piece of knowledge is found in different positions in the diagram under different social circumstances.

Therein lies the answer to a standard question about the laser case study. How is it, on the enculturational analysis, that people can now build lasers from instructions or even buy them off the shelf when once upon a time one could learn to build a TEA-laser only through social interaction with those who had the necessary skills? Building a laser is now like building a model airplane. To understand the change one must understand not only what we now know about lasers but what has happened to the users of laser-building knowledge.

On the one hand, scientific research on lasers has changed *short* to *8 inches*. This, as I have explained, has two advantages. First, it can take its place in the world of concerted behavior just like the "4" in "the mass of the neutrino is 4eV." Second, the cultural competence required to understand 8 inches, that is, the competence required to connect it to the rest of the world of electronics, is much more widespread among the population of potential laser builders than the cultural competence required to understand short. That is what we mean when, referring to figure 8.1, we say a piece of knowledge has moved upward.

Another change that has taken place simultaneously, however, is in the socialization of physicists. The upcoming generation will have learned more—purposefully and accidentally—during their enculturation about TEA-lasers and related problems. This must put them in a better position to understand and interpret explicit, written instructions than their predecessors. Perhaps they were taught more about TEA-lasers in their university degrees; perhaps they will have found it natural to chat over coffee with someone who knows the tricks of the trade and can show them the way to do things; perhaps they will have heard "war stories" that will have apprised them of the particular wrinkles involved in TEA-laser building; perhaps these things are just there in the social atmosphere.[10]

For the reasons explained in the last paragraph, written instructions that would not work once upon a time will work now. Serious scientific research has been involved but the increasing usability of laser knowledge cannot be understood without a grasp of the way the new knowledge relates to the foundations of culture in the lower two boxes in figure 8.1, how this is distributed, and how it changes over time. It is worth noting that unless you are one of the special group of physicists, *you* could still not build a TEA-laser from written instructions.

Finally, what is *an off-the-shelf* TEA-laser? It is a special kind of record-and-playback device. It is not an embodiment of laser-building skill but an embodiment of *the behavioral coordinates* of an act of laser-building skill.[11] For laser builders it will do the equivalent of mechanical paint spraying, with all the rigidity that is implied. It does not turn those who buy it into laser scientists any more than buying a calculator turns you into a mathematician. And, just as with the calculator the user had to fit its output into the world of arithmetic, so the user of a ready-made laser still has to fit it into the rest of the world of physics. (Would *you* know how to use an off-the-shelf TEA-laser?) For most of the purposes of most potential users, however, the mechanical device will be adequate.

Downward Movements

When knowledge moves upward on the chart it becomes available to the group of users who have learned the cultural competences belonging to science and arithmetic. Some of these will only be able to use the knowledge *as information,* after the manner of my *use* of "4" in "4eV." These people are less well off than those who can use it after the manner of lower level skills. In this respect the Dreyfus and Dreyfus model discussed in chapter 6 is illuminating. If one uses knowledge after the manner of use in the lower levels of the diagram, then one can use it more flexibly; there is a novice-like aspect to the use of knowledge as information. Of course, there are circumstances where this does not matter. For example, physicists who want to use a laser merely as a source of radiation will be happy using knowledge of the laser itself at the level of a novice. Sometimes, however, it is better for one's knowledge to move down to a lower category.

Go back to the strength of glass vessels. The laboratory technician who acted as domain expert on our crystal-growing project had first learned the strength of vessels from the chart shown in figure 8.2, but this chart was no longer used in the laboratory. As the technician put it, the knowledge had now entered the folklore of the physics department. Within the local culture of the department the knowledge had moved downward. In taking this journey from the public domain to the narrow location of the department, less restrictions were imposed upon its interpretation. It was now available in the form of *experience*; it was far more flexible and practically usable within that narrow location. To use the chart for crystal growing would require knowledge of the pressure that would build up inside the glass containers. To discover this, complex measurements and calculations concerning the chemical reactions taking place inside the vessels would be needed. This would not only be too costly and troublesome for single crystal-growing runs but would probably give the wrong answer. As we will see in later chapters, judging pressure inside these vessels was a matter of skill and experience that fitted well with the skill and experience required to judge the strength of the containers. To use the chart would have created the same sorts of inefficiency as a work to rule. That is why scientists discover they cannot do without their technicians; they cannot substitute formal knowledge for practical skills.

A downward movement also took place during the development of the TEA-laser. In the early, troubled days of the laser some exact figures for the tolerances allowable in the electrodes had been theoretically derived. Harrison obtained these figures and made the complex and difficult measurements needed to use them. Initially he was ready to try anything to make the laser work (Collins 1985, pp. 70–71). Later he discovered that this approach was unnecessary; eyeballing the configuration was enough. The use of his perceptual skills took just a few moments whereas the equivalent measurements took hours.[12] Once again, even though the science had developed to a point where the knowledge of electrode configuration was available at the top of the diagram, it was preferable to make use of a lower-level skill. Harrison, one might say, moved his knowledge of electrode tolerances down the diagram.

TEA-Laser Expert Systems: A Thought Experiment

As a thought experiment, imagine Bob Harrison trying to make his laser with the help of an expert system. How much use would the system be if it had been constructed at the various stages through which laser-building knowledge passed?[13] If the system had been designed at the putative apprentice-master stage, then there could be nothing in it about lead length however persistent the knowledge engineer who built it. At this stage knowledge about lead length was not consciously known and therefore could not be explicated by anyone. There is no special reason to suppose that inverted capacitors would come to figure in the system's heuristic rules because they might well be thought of as no more than a traditional part of the design—a decorative conceit. Most lasers built by properly trained apprentices would still work, but the crucial importance of lead length would remain unrecognized. Such an expert system would only work where there was already a substantial tradition of laser building among the users. Only users who already had a culturally induced propensity to invert their capacitors would find the system usable. In the language of figure 7.2, only users whose line b——b was high could use the device. Others would have to be referred to an expert should they either query top lead length or experience unexplained difficulty.

If, however, the system was built at a later stage, say around Harrison's starting point, then it could have included an instruction about keeping the top leads short. Nevertheless, this instruction would still only be usable by someone who was expert enough to know how to interpret the rule in the context of laser building. Cultural competence is still required but it is less cultural competence than would be required for success in the previous instance. Anyone who knew what *short* meant in the context of laser building would be able to use the system, whereas previously only those who had been around TEA-laser builders enough to find themselves impelled to make those strong steel frames for no particular reason would have succeeded. It may be that less competence is required by users at this stage but a naive laser builder, such as Harrison, who consulted an expert system on the matter, would still need an intervening expert to bridge the gap between the term *short* in the rule base and *short* as it should be understood by laser builders. Here then, the line a——a has come down a bit, but not enough to make contact with the b——b of a naive user. An intervening expert would still be needed.

Later, when electronics is discovered to be the relevant context, the rule can be made less ambiguous (for a particular design of laser). The instruction "keep the leads less than 8 inches in length" enables many more users with less specialized skills to succeed. To put this another way, the rule depends on far more widely distributed cultural competences; the meaning of "8 inches" is unambiguous for the native experts who live in Western societies because of the huge amount of technical socialization to which they have been subjected.

9

Explaining and Discovering Machines?

The theory developed in the previous chapters changes our understanding of the way computers contribute to the solution of problems. Using the theory I now look critically at three strong claims about the potential of computers. It is said that expert systems can explain their reasoning to the user after the fashion of an expert; it is said that knowledge engineers can solve the residual problems of expert systems by uncovering the *deep* knowledge pertaining to a domain; it is said that some computer programs have made scientific discoveries whereas others can induce new generalities from observations. How do these claims stand up to the social theory of machines?

Explanations in Expert Systems

I wrote PICKUP in MICRO-PROLOG, one of the programming languages that offers an explanation facility. At any time during the course of a consultation the user may ask "why" or "how"; the computer will then backtrack, copying to the screen the rules that prompted it to ask for information or to offer a particular conclusion. To give a simple example, PICKUP responds in the following way to a query about why it chose Albert as the person who would "come back to my place."

```
COMPUTER: ==> Albert
USER: why?
COMPUTER: To deduce
   Albert will-come-back-to-my-place
I used the rule
X will-come-back-to-my-place if
      X has-a-fleeting-smile and
      X will-maintain-prolonged-eye-contact and
      not X is-short-sighted
```

I can show
1 Albert has-a-fleeting-smile
2 Albert will-maintain-prolonged-eye-contact
3 not Albert is-short-sighted
Type a number
USER:
2
COMPUTER:
You told me that Albert will-maintain-prolonged-eye-contact

And so forth.

This is, of course, a ludicrously simple example. Explanations are as complex and ramified as the rule base of the system.

Explanation is a very attractive feature of expert systems. In the simple system for crystal growing, to be described in the next chapters, it was one of the features most appreciated by the users. Nevertheless, some unreasonable claims have been made for the power of automated explanation. As with the other uses of computers, to see what a computer can explain and what it cannot explain one has to think in terms of the relationship between what the system knows and what the user contributes.

The most ambitious claim for the power of explanation is that it gives expert systems moral superiority over other computers. Thus, Michie and Johnston (1985, p. 69) suggest:

> Any socially responsible design for a machine must make sure that its decisions are not only scrutable but *refutable*. That way the tyranny of machines can be avoided.

Not everyone makes quite such powerful claims for the potential of the explanatory facility of expert systems but the term "explanation" can easily be used as a "wishful mnemonic," to use Drew McDermott's (1981) phrase. An explanation can be number of different things.[1] The following explanation is entitled "What Is a Hologram?" I found it on a beer mat made for the Babycham Company in 1985.

> A hologram is like a 3 dimensional photograph—one you can look right into. In an ordinary snapshot, the picture you see is of an object viewed from one position by a camera in normal light.
>
> The difference with a hologram is that the object has been photographed in laser light, split to go all *around* the object. The result—a truly 3 dimensional picture!

This explanation is capable of making at least some people feel that they know more about holograms than they did before they read it. The words on the beer mat are not simply nonsense nor would we be likely to mistake them for, say, a riddle or a joke or an advertisement. Presumably there are people now alive in England who have studied the beer mat and, if asked, "Do you know how a hologram works?" would reply, "Yes," whereas immediately before they had read the beer mat they would have answered, "No," to the same question.

The beer mat example is extreme but it has much in common with other explanations. For example, my understanding of general relativity is rather like the understanding of holography I could obtain by reading the beer mat. This in turn is like my understanding of the rest mass of the neutrino. The explanation of a hologram found on the beer mat, like my knowledge of the rest mass of a neutrino, does not put me in a better position to act. For example, it does not put me in a position to make a better hologram nor does it put me in a position to *refute* anything that I could not refute before. Of course, it might make me *feel* as though I understand holograms better than before.

Turning to expert systems, they explain by backtracking through the tree of rules that brought them to the point of enquiry.[2] The very best that such a system can do is to convey to the user everything that the expert has put into the program. Think in terms of three models of knowledge: the Expert's Model, which is everything the expert knows; the Encoded Model, which is that part of the Expert's Model that can be encoded in the program; and the User's Model, which is what the user knows. Assuming the user can read, by assiduous deductive reasoning he or she can reconstruct the Encoded Model of the domain by using the system's explanations. Potentially, the user can recover the whole of the rule base that drives the systems.

This is no small thing. In the crystal-growing expert system we found that users loved to explore the rule base in this way; they felt that it would be a very useful teaching device. Nevertheless, mastering a knowledge domain means much more than learning the Encoded Model. The rule base alone can provide knowledge only as deep as beer-mat knowledge, which is useful only for the performance of machine-like acts. Expert systems' explanations alone do not put the user in a position to *refute* or *controvert* the system's reasoning. (They do put the user in a position to spot errors of logic or programming mistakes, but these are not my concern.)

Who can controvert the reasoning of a logically coherent expert system? The answer is that decisions based on the Encoded Model can be controverted from the Expert's Model or any other model that contains relevant aspects of tacit knowledge not available to the machine. Scrutability of decisions can lead to refutability only if the scrutineer knows more than the machine; that is, if the User's Model of the domain approaches the Expert's Model in terms of its cultural depth. We are back with spying in Semipalatinsk; only the native can refute the trained spy's account of what it is to be a native.

To see how this works consider a familiar example, the English language. Imagine an expert system for the domain of spoken English. This is not an expert system with a natural language interface but a system in which the expertise comprises rules for the production of English. Ignore deep problems to do with whether or not such a machine could actually be built and what type of English it would speak. Now imagine certain explanatory interactions with this program. For example, imagine the program produces the sentence: "It cold is," and imagine the user asks, "Why?" and that the program replies:

> I used the rule "expressions with verbs may be English sentences and the verb goes at the end in winter."
>
> "It cold is" has a verb and "cold" suggests winter, therefore this is a proper sentence of English.

Now imagine this explanation being encountered by a range of users.

First, let it be encountered by the English-speaking knowledge engineer who designed the system. In this case it could be used to diagnose an inadequacy in the rule base.

Second, let it be encountered by a native English speaker who was not a knowledge engineer. It could then be used to controvert the system's advice regarding the nature of English sentences; Michie and Johnston's ambition can be fulfilled in this case because here the user would draw on an Expert's Model of English to controvert the Encoded Model.

Third, imagine it being encountered by someone with a beginner's acquaintanceship with English such as might be acquired from an elementary primer without conversational practice. In such a case there is no reason why the explanation should not convince the user. Even if the user had enough previous English book-learning to be suspicious enough to ask, "Why," the explanation would dispel residual doubt. This type of novice user would think his or her grasp of idiom was becoming deeper as this new rule was believed and absorbed. Unfortunately, they would be mistaken—a dupe of the tyranny of machines *because* of the explanatory facility provided.

Fourth, imagine the program being encountered by someone who was just beginning to learn English. In this case the user would learn a new rule—as it happens, an incorrect one.

Fifth, note that speakers who felt that they were accomplished in the language, but had learned all their English by using the system as a teacher, would never be in a position to reject any rule provided by the system other than those arising from logical inconsistencies.

The simple example shows that treating an expert system as an isolated intelligence is quite unhelpful in determining what is going on when it tries to explain itself. We must know who is using it and what they know.

With this in mind, consider some other ambitions for the explanatory powers of expert systems. In the following discussion the quotations are taken from papers presented at a conference on the topic "Explanation in Expert Systems."[3]

> There are two compelling reasons for the requirement that expert systems should be able to explain their reasoning and justify their conclusions . .
>
> (i) The consumers of automatic advice need to be convinced that the reasoning behind a conclusion is substantially correct, and that the solution proposed is appropriate to their particular case.
>
> (ii) The engineers of an expert system need to be able to satisfy themselves that the mechanisms employed in the derivation of a conclusion are functioning according to specification. (Jackson 1986)

Ambition (i) is likely to be dangerously easy to fulfill in the case of naive users because the explanation facility will simply make the machine look clever. The explanation will be a device for cutting off questioning with a dazzling display of knowledge. "Blinding with science" is the appropriate phrase. It seems wrong to expose users to this. It is, however, a proper aim for systems designed for use by *experts*. The naive will need to be protected.

Ambition (ii) is, of course, entirely appropriate because the system's Encoded Model is to be compared with an Expert's Model.

Whilst explanations are seen as important in allowing the user to "validate the program's reasoning" (Shortliffe et al. 1975), and would be useful to learners exploring the system in order to acquire expertise themselves (Shortliffe et al. 1975; McKeown 1980), it is in the possible use of expert systems in a tutorial capacity as proposed for example by Clancey (1984), that effective explanations may be particularly crucial. (Rymaszewski 1986)

This aim, the use of expert system explanations in a tutorial capacity, is a good idea so long as the users are being given a grounding by standard apprenticeship methods in addition to the information they get from the system. Otherwise they will gain only certain aspects of knowledge.

To genuinely answer a user's request for explanation might sometimes be too difficult even for a human expert and could give the user information overload. Here a limp attempt at explanation, a reference to a text, a mere illustration, an admission of inability or conceivably even downright evasion might be appropriate. (Dodson 1986)

Although admitting incompetence and redirecting the user elsewhere is a very sensible policy for expert systems, it is hardly explanation. On the other hand downright evasion—easily managed with beer-mat-type explanations—will reinforce rather than reduce the tyranny of machines.

I propose that explanation be considered as the process of conveying understanding. A similar but more restrictive sentiment is expressed by Jackson and Lefevre (1984) when they describe an expert system's explanation as the process of communicating and understanding of what the system itself does. (Hughes 1986)

Unfortunately, conveying understanding cannot be done by expert systems in anything but the most superficial sense. For lay users explanation is more likely to mislead than to aid understanding of what the system is capable of.

The system will need to support the user in developing a deep level of understanding. One important aspect of this approach is that explanations should be offered and also allowed to progress though out the interaction while gradually building the user's conceptual framework and overall understanding of the inter-related problems that have to be dealt with . . . the advice system will need to construct arguments dynamically if it is to do more than just *enhance the user's knowledge and understanding.* There is a need for the user to gain both factual information and procedural information in order to be able to make sensible actions, i.e. the user will have to know what to do and how and when to do it. Consequently the second aim of the system is to equip the user with *what to do next* and *how to do it.* . . . the advice system shares many of the problems that are inherent within intelligent teaching system research and as with such work there is a need for the advice system to maintain a two way interactive-learning process where both system and user learn about each other in a dynamic manner. (Crossfield 1986)

Whether the system can "support the user in developing a deep level of understanding" depends on what else is going on in the way of teaching. A system on its own can never give a lay user anything but the superficial knowledge of the Encoded Model.

The importance of a separate explanation facility is highlighted by the work of Evans and Wason (1976) showing that people exhibit a remarkable propensity to produce justifications of a solution even when the solution is wrong. They suggest that the rationalization of a decision does not necessarily correspond to the actual thought processes that took place in arriving at the decision. If this is so, then the Assistant must know not only how the Expert actually reached its decision, but also how best to explain the decision. This will often involve introducing knowledge that did not actually figure in the reasoning process. (Knight 1986)

This aim seems to involve providing users with rationalizations. Again this can only add to the tyranny of machines where naive users are concerned even if it is but a reflection of the existing tyranny of human experts; it will prove irritating where the users are experts.

Boden (1985) suggests that it is a positively bad thing that intelligent computers are attaining natural conversational competence because this enables them to give the impression that they are more expert and less fallible than they really are. Clearly the same sort of dangers may be found in the explanation facilities of expert systems when they confront naive users. The major point, however, is that we cannot understand explanation in computers unless we think of them as replacing humans in interactions. However sophisticated a "model of the user" the system can generate, existing programs can still only make good deficiencies in the user's *information*; they cannot provide the sort of understanding of a knowledge domain that is normally acquired through apprenticeship or socialization.

Deep Knowledge

Understanding that is normally acquired through socialization was referred to in the last chapter as "low level knowledge." This term reflects the position of such knowledge on the hierarchy represented in figure 8.1. Expert systems cannot encode low level knowledge. The problem is sometimes resolved by the movement of knowledge up the hierarchy into the more accessible boxes. In the last chapter I explained some of the mechanisms involved in this process.

It seems to be the ambition of some knowledge engineers to do for themselves the scientific research that moves knowledge upward (though they don't think of what they are doing in these terms). It has been claimed that the endless complexity of common-sense rules encountered by knowledge engineers can be simplified by uncovering the "deep" knowledge of a domain.[4] The best that knowledge engineers can expect, however, is to help clarify the logic of a knowledge domain. One can see why by considering once more the early days of the TEA-laser and imagining a knowledge engineer trying to build an expert system to cope with that problem.

Assume that the knowledge engineer is not a better physicist than the laser scientists themselves and is not in a position to make the relevant discoveries in electronics that led to the resolution of problem of the top lead. Nevertheless, the knowledge engineer, by assiduous study of what TEA-laser physicists do, or by using some technique for probing the cognitive structures that underlie the way experts see the world, might discover that all the successful experts were, as a matter of fact, using top leads that were less than 8 inches in length; it was just that no one had noticed it. This is not impossible and would represent a very important role for the knowledge engineer. Indeed it may seem from my account that any astute observer would be likely to notice such a thing, especially if assisted by a battery of psychologically sophisticated questions. In reality it is extraordinarily unlikely.

The reason it is so unlikely is that when scientific research is done in real time there is too much going on. Thus, when the laser was being built it was not at all clear that leads were important because there were an indefinite number of other differences between successful and unsuccessful lasers competing for attention. There were matters of voltages, inductances, rise times and pulse shapes of the charge, geometry of the electrodes, separation of the trigger wires, number of trigger wires, surface smoothness of the electrodes, exact constituents of the gas in the laser cavity, rate of change of gas, cavity material, humidity, temperature, electrical characteristics of the laboratory, spatial arrangement of components both within and without the laser, and so on and so on.[5] That lead length was ever picked out for special attention had to depend at some point on theoretical appreciation of the laser, but even that was far from sufficient. The point is that noticing lead length was a matter of physics not common sense or psychological sophistication.

The role of the physics is not obvious because of a fundamental misunderstanding of the process of scientific research. The misunderstanding arises because stories of scientific research are nearly always told retrospectively. It is only because I have told the story of the TEA-laser retrospectively that the top leads have salience. At the time everything seemed equally salient. This sharp and crucial difference between real time experience and retrospective accounts has been discussed earlier. It is the difference between bringing actions under a rule retrospectively and predicting the future. We can backhack to solve problems of artificial intelligence but we cannot forehack. We can have a backward-looking causal social science but not a deterministically predictive social science. We can see all our scientific mistakes when we look backwards, and we can see how to solve them but this does not help with today's science. We know how the universe was, but this is not the same as finding out how the universe is.

The same fundamental error underlies two related issues in AI and expert systems: automated discovery and machine induction.

Automated Discovery

There is a program called BACON, written by Langley, Simon, and others (see Langley et al. 1987). This program is supposed to be capable of rediscovering fundamental laws such as Boyle's law and Kepler's law. The program is supposed to achieve this by taking "experimental results" and inducing the laws from them via a problem-solving algorithm. Once more the error lies in the difference between real science and retrospective science. Given clean data *that already conform* to the expected result there is no reason why a program should not uncover the underlying relationship. But suppose some of the readings contained errors—as all real science does—how would the program cope?

Many detailed studies of scientific discovery have shown that the inspiration lies in the sifting of the data allied with faith in the way the world works rather than in inducing a rule from fixed data points. I think it is true to say that in all cases of crucial experiments so far analyzed in historical detail the raw data would be incapable of distinguishing between rival hypotheses. Only in retrospect does this not appear to be the case. Certainly this can be said of Boyle's experiments on the vacuum (Shapin and Schaffer 1987); Millikan's discovery of the unitary charge on the electron (Holton 1978); of the Michelson-Morley experiments on ether drift (Polanyi 1958); of Eddington's observations of bending of light rays by the sun, which proved general relativity (Earman and Glymour 1980); of gravitational radiation experiments (Collins 1985); and so on.[6] Again, a simple example reveals the nature of the problem. Though what follows is not a report of real scientific work it does encapsulate the problems of real science at the frontiers of experience.[7]

School students of mathematics were asked to determine the relationship between the sides of a certain geometrical figure. They were asked to draw arbitrarily sized versions of the figure and then to measure the sides. By combining the raw data, two pairs of columns of numbers were computed, one by the students and one by their teacher (see table 9.1). Each row of numbers shows the results of the students' and teacher's computations for one size of figure. The pairs of columns can be seen as tests of competing hypotheses about the relationship between the sides of the figure. A numerical match between column A1 and column B1 supports the students' hypothesis, a mismatch reduces its credibility. In the case of columns A2 and B2 it is the teacher's hypothesis that is at stake.

For the reader the problem of deciding which of the columns best supports the hypothesis is something like a real scientific problem in so far as the numbers themselves are ambiguous. For example, although the results for Drawing 14 give more support to the students' hypothesis, Drawing 13 favors the teacher. Statistics might help to determine which row best fits its underlying hypothesis on the assumption of randomly distributed error but, given

Table 9.1
Competing geometrical hypotheses

	Students' Hypothesis		Teacher's Hypothesis	
	A1	B1	A2	B2
Drawing 1	131.4	128.4	16.1	15.95
Drawing 2	74.0	70.6	12.0	11.6
Drawing 3	134.7	130.0	16.4	16.1
Drawing 4	39.7	37.7	8.8	8.15
Drawing 5	126.8	130.8	15.9	16.1
Drawing 6	100.0	98.0	14.0	13.85
Drawing 7	157.2	159.3	17.7	17.9
Drawing 8	192.3	190.3	19.6	19.7
Drawing 9	126.7	123.3	15.9	15.65
Drawing 10	151.9	150.5	17.4	17.45
Drawing 11	242.2	245.5	22.0	22.55
Drawing 12	200.3	205.7	20.0	20.45
Drawing 13	194.7	190.0	19.7	19.7
Drawing 14	87.9	88.0	13.0	13.55

nonrandom experimental error, it is not possible to determine which is correct from the numbers.

Now, the left-hand columns are the outcome of a standard, real-life, classroom demonstration of the Pythagorean theorem. The figures the students were asked to draw were right-angled triangles; they were asked to measure their sides and compute their squares and relevant sums. In the test of the students' hypothesis, column A1 is the result of calculating a^2+b^2, whereas column B1 is the result of calculating c^2. The correspondence between the numbers in the two columns is taken in the classroom to demonstrate the Pythagorean theorem.

The right-hand columns are also based on the students' measurements but are not usually part of the exercise. The teacher calculated a+b to produce column A2, whereas the numbers in column B2 represent 1.5c-1.[8] The crucial difference between a retrospective demonstration and the same "experiment" seen as a real-time determination of a mathematical relationship is easily seen. The data can be taken to demonstrate the Pythagorean theorem only after the universe of possibilities has been narrowed to the binary choice between Pythagorean and non-Pythagorean—all competitors such as a+b=1.5c-1 must already have been eliminated. In real life and real time such processes happen within the social collectivity. Machines cannot reproduce them.[9]

Machine Induction

The same problem arises for so-called machine induction as used in expert systems. The idea here is for the computer to be given a set of examples rather than rules and then induce the underlying relationships for itself.

If applied AI is to be important in the decades to come . . . we must develop more automatic means for what is currently a very tedious, time-consuming and expensive procedure. (Feigenbaum and McCorduck 1984, p. 85)

Machine induction is really another way of making automated scientific discoveries though the algorithms are different from BACON's, and the examples can be entered in the form of words rather than numbers. It has been shown that what the machine will induce is crucially dependent on the examples and that additional examples do not necessarily refine the rule base in an incremental way; small increments may produce catastrophic revisions to the induced rules.

The methods used are still immature and the rules theretore lack robustness. For example, they are often rewritten completely if one new case is added to the database." (Kidd and Welbank 1984, p. 78)

This problem is almost certainly not a matter of immaturity of methods. It arises from a fundamental difficulty (see Bloomfield 1986). Again a well-worn imaginary example will make the point clear. Imagine the machine trying to induce an underlying relationship from the following numbers:

2, 4, 6, 8

A reasonable first supposition is that the rule is "add two" to the previous number.

Now consider the addition of one additional term to the sequence:

2, 4, 6, 8, Who

The underlying relationship has changed catastrophically from being a number series to being part of a song.[10] As with automated inductive reasoning for scientific discovery, data drastically underdetermine predictive generalizations.[11]

As my argument has shown, the idea of trying to solve the problem of knowledge elicitation by substituting machine for human induction is mistaken.[12] It is induction that is the ultimate problem; it is induction that we can do and that machines cannot. Given modern analyses of the resolution of scientific controversies it is extraordinarily unlikely that machine induction would converge with human induction. To think of automated induction as a *solution* to the problem of induction is a strange reversal of the order of things.

III

A Skill Encoded—A Skill Practiced

10

Tales from the World of Crystal Growing

Experimental semiconductor crystals are grown in the physics laboratories of the University of Bath. Enlisting the help of two colleagues, I set out to develop an expert system for crystal growers. The experience illustrates most of the themes discussed in the earlier chapters. The major contrast is between the endlessly ramifying description of the actual practice of crystal growing, a short example of which is presented in chapters 11 and 12, and the schematic encodings found in textbooks, experts' talk, and computer programs. These are described in this chapter. Other themes will be highlighted as they occur throughout the three chapters.

The Project

We began the project in 1984. Bob Draper, a technician in the physics department of the university, and Rodney Green, from the university's School of Management, agreed to help with the project. Draper agreed to be the subject of prolonged questioning by Green and to take me through a short, on-the-job training in the rudiments of growing crystals of experimental semiconductor materials. Green had already published in the area of expert systems (Green and Wood 1983) and was helping a number of firms develop systems of their own for use in manufacturing. Green agreed to elicit Draper's knowledge with a view to developing the expert system. I would look on and record this process while simultaneously undergoing my crystal-growing apprenticeship. The object was for me to make a comparative study of knowledge elicitation and hands-on learning.

Because of the comparative aspect of the project, Green's elicitation sessions were unusually restricted. To keep the apprenticeship and elicitation separate, Green was allowed only one initial tour of the laboratory. Thereafter all his interchanges with Draper took place in his office. He was able to read and talk about crystal growing but not to see the process in action. The idea was for me to get a clear idea of just what could be learned from reading and asking as opposed to seeing and doing. Normally, knowledge elicitation would use a far wider range of techniques, but the object of this exercise was to understand the separate contribution of these different inputs. Green's "mistakes" should be understood in this light; they were partly a consequence of the experimental design.

The experiment is represented diagrammatically in figure 10.1.

The initial plan was that Green alone would be responsible for turning the knowledge he elicited from Draper into an expert system. In the end contingencies prevented him from making more than a start and I completed the system. It was called CRYSTAL and was written in MICRO-PROLOG utilizing APES (Augmented Prolog for Expert Systems).

The three of us first met on March 18, 1984. We looked round the physics laboratory with Draper on March 21.[1] After his tour of the laboratory Green began his study of the area by reading a standard volume on the subject (Pamplin 1980).[2] It is important to note that this was the reference book regularly used by Draper. He kept a copy on his shelf and recommended it to us.

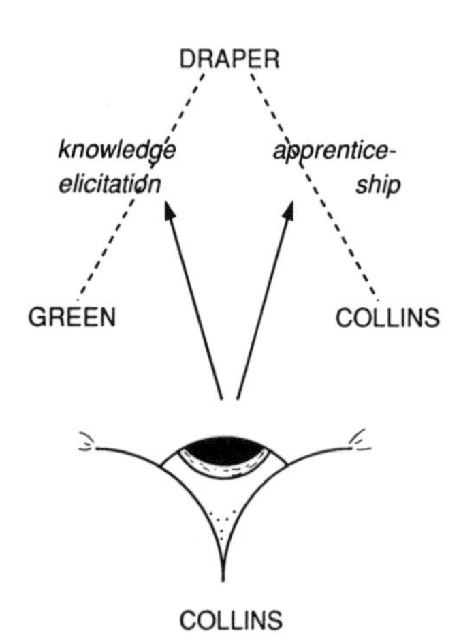

Figure 10.1
The experimental design

Descriptions of Crystal Growing

I want to compare different descriptions of crystal growing. From a theoretical point of view the best place to start would be with the textbook, but for noncrystal growers this would be boring if not incomprehensible. To introduce the subject in an easy way I will use the schema that formed the backbone of CRYSTAL's rules. This schema was wrested from the knowledge elicitation sessions. Just how this was done is described in the last part of the chapter. As will be seen the schema differs from the textbook account and from the practice of crystal growing in both style *and content.*

Introduction to Crystal Growing at Bath

In Bath two main methods are used for growing crystals. The main difference is in the means of containment of the materials. In one method, *zone melting,* the raw materials are contained in an open *boat* that lies horizontally. A narrow radiant heater moves from end to end of the boat, melting a zone in the material that resolidifies as the heater passes on. The procedure may be carried out a number of times. Alternatively the raw materials may be contained in a closed *ampule* that is heated in a vertical position and then cooled very slowly from the bottom. The freeze front moves slowly upward through the material. This method was always referred to by us in conversation as *gradient freezing,* but it is also known as normal freezing or directional freezing. In both zone melting and gradient freezing one first melts then freezes the material, and both come under the generic heading of melt-growth.

For very high melting point or very high vapor pressure substances a variant of the ampule method called *solution growth* may be used. In solution growth the substances are dissolved in a relatively low melting point solvent, such as tin, and then allowed to crystallize out of solution. The problem is to get rid of the solvent at the end of the process. For example, tin must be dissolved by a suitable acid so as to leave the crystals free but unharmed. Solution growth is a "dirty" and slow method.

Zone melting at Bath is done at atmospheric pressure. This means that the open boat method cannot be used when constituents include materials that evaporate or boil at the highest temperature that the process will require—usually the melting point of the least volatile constituent. At these high temperatures the volatile constituents will escape from the mixture before the crystal is formed. For these *high vapor pressure* elements the ampule method is used. A thick-walled closed ampule can resist a high internal pressure and contain volatile elements. The strength of an ampule depends on its bore and wall thickness.

Note the following two points carefully as we will return to them: certain elements—such as phosphorous—have such high vapor pressures at the melting points of other constituent elements that the ampule method becomes impracticable; solution growth may be used to resolve this problem. The ampule method cannot be used to grow crystalline substances that expand on freezing; the expansion will crack the ampule and damage the crystals as the melt solidifies.

A number of factors affect the choice of material from which boats or ampules are made: boats and ampules must not melt at temperatures below or near to the melting temperature of the constituents; they must be chemically inert to the constituents.

CRYSTAL

All the above knowledge was encoded as rules in CRYSTAL along with some specific details about melting points and strengths of containers. Because I have already described the rules in words it seems sensible to present CRYSTAL now even though it was not completed until late in the project. The "intelligent" part of CRYSTAL looked as in figure 10.2. It is worth reading some of this program to get the feel of this condensed representation of crystal-growing knowledge.

A consultation with CRYSTAL looks as follows:

find(X:X OK(Tin-Arsenic))
MOST CRYSTALS DO NOT UNDERGO A PHASE CHANGE BELOW THEIR MELTING POINT. IS THAT ALSO TRUE FOR ((TIN-ARSENIC)) ?
yes
WHAT IS THE APPROXIMATE MAXIMUM PRESSURE IN ATMOSPHERES WHICH YOU WOULD EXPECT THE REACTION INVOLVING (TIN-ARSENIC) TO REACH IF GROWN BY STRAIGHT-FORWARD METHODS ?
ANSWER IS 3
[This number, for reasons which will become clear, is only a guess, but it does reflect the volatility of arsenic.]
IS IT TRUE THAT (TIN-ARSENIC) EXPANDS-ON-FREEZING ?
no
WHAT IS THE MELTING POINT OF (TIN-ARSENIC) IN DEGREES ?
ANSWER IS 231
IS IT TRUE THAT (TIN-ARESENIC) CONTAINS-METAL ?

yes
IS IT TRUE THAT (TIN-ARSENIC) CONTAINS-AN-OXIDE ?
no
==> (FREEZE-GROWING IN AMPULE-WITH-40MM-OR-LESS-EXTERNAL-DIAMETER-AND-1MM-WALLS MADE FROM PYREX)

CRYSTAL contained general rules, but it also contained some quite specific information about the strength of glass tubes. As mentioned in chapter 8, I encoded this information using the glass-blower's chart (figure 8.2). But the chart contains much more information than the few possibilities listed incrystal (see the rules about ampule strength in figure 10.2). The reason I used those particular values was that tubes of those sizes were available in the laboratory at Bath. I simply measured what was on the racks and encoded the corresponding values. This enabled the system to prescribe dimensions of tubes that could be found in the laboratory without the user having to know anything about the strength of glass tubes or being able to read the chart. In this way CRYSTAL could do a little more for an ignorant user than even the glass-blower's chart. In principle this aspect of locally applicable complexity of the rule base did lower the line b ——b (figure 7.2). In this sense it did, in principle, replace a little bit of expertise by relating glass-blowing knowledge to much more widely distributed abilities. But, of course, this was bought at the expense of a limited flexibility. The knowledge could only be used in the Bath laboratory and only so long as the stock of tubes was maintained in its current form. (As we will see, the information was in practice useless, but for different reasons.)

CRYSTAL expressed its knowledge in a few rules which, because of their brevity and apparent authority, engendered confidence in the user. In this respect the system was like a textbook. CRYSTAL, however, differed quite markedly from the textbook in terms of content. The rules included rules of thumb extracted, with some difficulty, from our expert advisor. The textbook presented an altogether more theorized version of crystal growing as the following extracts show.

Pamplin (1980)

Astonishingly, Pamplin (1980) contains absolutely nothing about the use of closed ampules as a means of containing high pressure reactions. In this it is a typical text. According to our expert

```
(X|Y) wont-melt-container-material Z if
          (X|Y) has-melting-point x and
          Z will-soften-at y and
          x is less than y

X wont-be-dissolved-by-or-reduce (Y|Z) if
          not (Y|Z) dissolves-container-material X and
          not (Y|Z) will-be-reduced-by X

(X|Y) dissolves-container-material Z if
          (X|Y) contains-metal and
          Z is-a-metal

(X|Y) will-be-reduced-by Z if
          (X|Y) contains-an-oxide and
          Z is-a-reducing-agent

pyrex will-soften-at 501
quartz will-soften-at 1251
carbon will-soften-at 3001
ceramic will-soften-at 1701
platinum will-soften-at 1901

(X|Y) wont-damage-or-escape-from-the-container Z if
          the reaction involving (X|Y) will reach no more than x atmospheres and
          Z will withstand reaction pressure of y atmospheres and
          x is less than y and
          not (X|Y) will-crack-the-container Z

boat will withstand reaction pressure of 1.01 atmospheres

ampoule-with-40mm-or-less-external-diameter-and-1mm-walls will withstand
        reaction pressure of 3.01 atmospheres
ampoule-with-27mm-external-diameter-or-less-and-1mm-walls will withstand
        reaction pressure of 4.01 atmospheres
ampoule-with-32mm-or-less-external-diameter-and-2mm-walls will withstand
        reaction pressure of 7.01 atmospheres
ampoule-with-20mm-or-less-external-diameter-and-1mm-walls will withstand
        reaction pressure of 8.01 atmospheres
ampoule-with-14mm-or-less-external-diameter-and-1mm-walls will withstand
        reaction pressure of 9.01 atmospheres.
ampoule-with-20mm-or-less-external-diameter-and-2mm-walls will withstand
        reaction pressure of 12.01 atmospheres
ampoule-with-09mm-or-less-external-diameter-and-1mm-walls will withstand
        reaction pressure of 14.01 atmospheres
ampoule-with-06mm-or-less-external-diameter-and-1mm-walls will withstand
        reaction pressure of 20.01 atmospheres
ampoule-with-27mm-or-less-external-diameter-and-4mm-walls will withstand
        reaction pressure of 22.01 atmospheres
ampoule-with-12mm-or-less-external-diameter-and-2mm-walls will withstand
        reaction pressure of 25.01 atmospheres
ampoule-with-09mm-or-less-external-diameter-and-2mm-walls will withstand
        reaction pressure of 30.01 atmospheres
```

```
zone-refining is-right-method-for-container X if
          X is-a-boat
freeze-growing is-right-method-for-container X if
          X is-an-ampoule

boat is-a-boat

platinum is-a-metal

carbon is-a-reducing-agent

(try partial-compounding or try solution-growth The method and container will depend
         on the partial compounds or on the solvent) OK (X|Y) if
          not Z is-one-straightforward-method-for-growing-crystals-of (X|Y) and
          (X|Y) does-not-have-a-phase-change-below-its-melting-point

try-solution-growth-at-temperature-below-phase-change-point OK (X|Y) if
          not (X|Y) does-not-have-a-phase-change-below-its-melting-point

X OK (Y|Z) if
          X is-one-straightforward-method-for-growing-crystals-of (Y|Z)

(X|Y) can-be-contained-by (Z made from x) if
          (X|Y) wont-damage-or-escape-from-the-container Z and
          (X|Y) wont-melt-container-material x and
          not Z cant-be-made-from x

(X|Y) will-crack-the-container Z if
          (X|Y) expands-on-freezing and
          Z is-an-ampoule

(X in Y made from Z) is-one-straightforward-method-for-growing-crystals-of (x|y) if
          (x|y) does-not-have-a-phase-change-below-its-melting-point and
          (x|y) can-be-contained-by (Y made from Z) and
          X is-right-method-for-container Y and
          Z wont-be-dissolved-by-or-reduce (x|y)

X is-an-ampoule if
          X will withstand reaction pressure of Y atmospheres and
          not Y is less than 1.2

X cant-be-made-from Y if
          X is-an-ampoule and
          Y cant-be-made-into-an-ampoule

platinum cant-be-made-into-an-ampoule
carbon cant-be-made-into-an-ampoule
ceramic cant-be-made-into-an-ampoule
```

informant one will not find the major division of methods as described in my "Introduction to Crystal Growing at Bath," and as encoded in CRYSTAL, in any texts. In Pamplin (1980) high pressure reactions are treated as problems to be solved by building special pressurized apparatus. "Most textbooks go for complicated solutions to simple problems," according to Draper. Of course, pressurized apparatus offers a general solution to the problems—certainly it is the only sensible way of growing large numbers of crystals in a controlled way—but it is not the way to grow a cheap experimental crystal quickly. It is completely inappropriate as a local solution.

Pamplin (1980) opens with a diagram that maps out the universe of crystal-growing methods (see figure 10.3). This diagram will play a significant part in what follows.

The diagram shows a large variety of methods, some of which are meant to cope with the problems of awkward materials with high vapor pressures and high melting points. (The horizontal boat and ampule methods are referred to in the diagram as zone melting and normal freezing—they are in the center of the penultimate column.) In his introductory chapter Pamplin (1980, pp. 6–7) treats these methods as follows:

> 1.4 Melt Growth Techniques
>
> 1.4.1. Introduction
>
> . . . melt growth normally requires that the material melt congruently (that is it does not decompose below or near its melting point) and has a manageable vapor pressure at its melting point. Thus a great many materials cannot be grown from the melt. . . .
>
> 1.4.2. Normal Freezing, Directional Freezing, or the Bridgman-Stockbarger Method
>
> The most straightforward and inexpensive melt growth technique is normal freezing—a molten ingot is gradually frozen from one end to the other. . . . The usual configuration is vertical with the melt in an ampule being lowered slowly from the hot zone to the cooler zone which is below the melting point. . . .
>
> If the material expands on freezing and would crack the vertical ampule, or if it is desirable to use a graphite or vitreous carbon boat, horizontal normal freezing is used. This is also the more convenient configuration when the vapor pressure of a component is to be controlled by either liquid encapsulation or having a connection to a source at a fixed temperature.

Though there is further discussion of these methods in later chapters and a whole chapter on "Methods of Growing Crystals

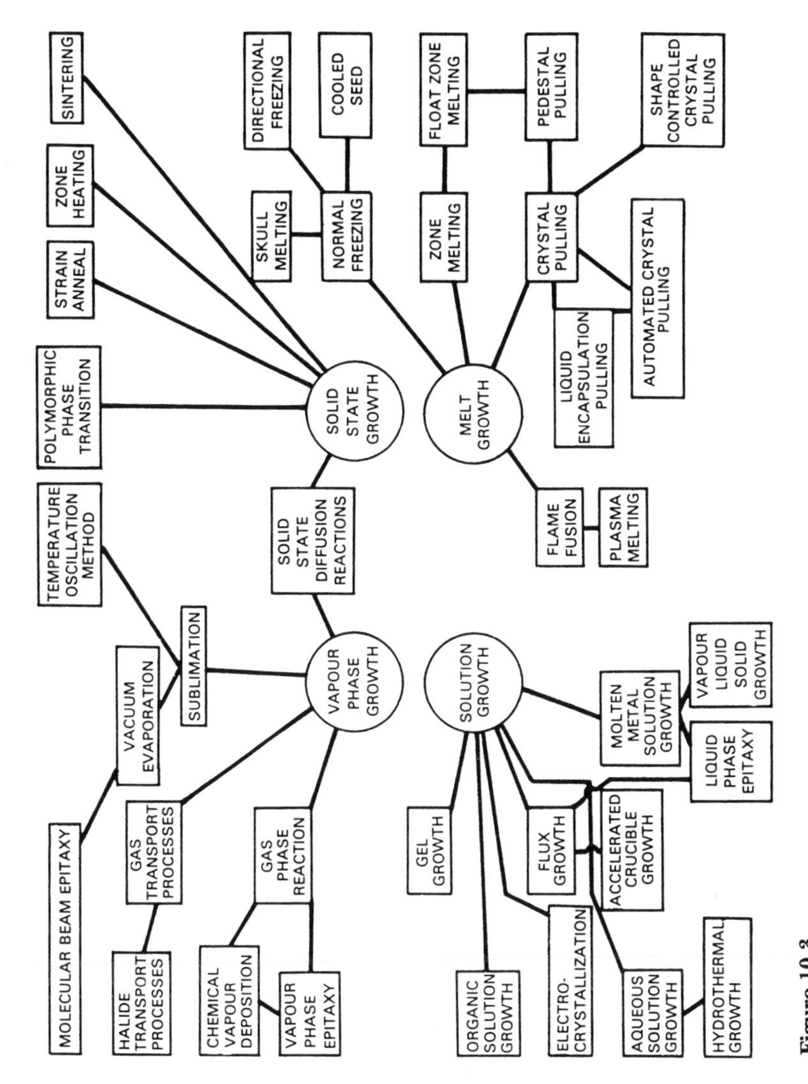

Figure 10.3
Crystal growth techniques

Under Pressure," nowhere does the closed ampule figure as a simple pragmatic solution to the *pressure* problem.

It is worth noting, for reasons that will become clear, that Pamplin (1980, p. 9) favors "crystal pulling," or the Czochralski technique:

1.4.4. Crystal Pulling

Crystal pulling dates from Czochralski's work on the speed of crystallization of metals published in 1918. Its real importance, however, starts in the early 1950's when Teal and Little at Bell Labs developed it for the production of pure and doped crystals of germanium and silicon and even grew pn junctions. It has subsequently been used for some III V semiconductors using, when there is a high vapor pressure of arsenic or phosphorous, the liquid encapsulation technique ("LEC pulling"). In the early 1960's pulling in air was pioneered for many laser host materials like $CaWO_4$.

And a little later:

1.7 Choosing a Crystal Growth Method

For bulk growth of high quality single crystal material seeded melt growth (e.g. crystal pulling or float zone melting) is undoubtedly the best method available today for congruently melting materials. As in the cases of Si, GaAs, and GGG it is fast, efficient, and can be automated, and it produces the most perfect crystals possible.

Note the style and universalistic authority of the text, the way it cites sources and leads inexorably to a rationally chosen best method. Compare this with what follows.

Knowledge Elicitation and Crystal Growing

I mention the emphasis on crystal pulling in the text because it illustrates the way the book influenced the knowledge elicitation sessions between Green and Draper. In spite of its strong recommendation in Pamplin's book, crystal pulling is not a method regularly used at Bath. Nevertheless it was the topic of more discussion between Green and Draper than any other method. This is how the topic was introduced by Draper to Green during their conversations.

Green: [referring to the chart of crystal growing methods found at the front of Pamplin's book, figure 10.3] Is one of these by far and away the most convenient to use if it's possible?

Draper: The most convenient to use is, in general, the crystal-pulling method because it will give you large crystals of the orientation that you want in a fairly short space of time.

After some discussion of crystal pulling, Draper pointed out that simple apparatus could only handle pressures up to about 1.2 atmospheres. Above this pressure an expensive pressurized apparatus and safety room was required. The "six-figure" cost of this was beyond the means of most universities. The discussion of atmospheric pressure Czochralski method continued:

Green: [seeking the first rule for the expert system] The essential rule is: if the vapor pressure is less than 1.2 atmospheres, then the method to use is crystal pulling. . . . What's wrong with that?

Draper answered that the major problem, even for a low pressure apparatus, was cost. An ordinary crystal puller cost about £30,000. Also there were problems even for some low vapor pressure mixtures, which required the technique of "Liquid Encapsulation Pulling" (see figure 10.3). In this technique a layer of viscous molten encapsulant (almost invariably boric oxide) floats on the surface of the mixture preventing evaporation while the crystal is pulled through the liquid surface. The encapsulant must be dissolved away later. Problems arise if the melting point of the main mixture is too low to allow the encapsulant to melt. (As we will see, this rules out its use for bismuth-arsenic combinations.)

Only after more than two hours' discussion of crystal pulling and its variants did a picture begin to emerge that more closely related to Draper's experience. This happened when Green tried to move on to what he hoped would be the second choice method—zone melting.

Green: Zone melting. When would you use that?

Draper: In most cases because it's probably the easiest of all.

Green and Collins: [laughter at the sudden realization that this seemed to contradict the previous hours of conversation.]

Draper: It is a very easy process to set up and it is a very common one.

Green: How would you choose to use zone melting rather than crystal pulling?

Draper: Cheapness—you can build a zone refiner for thirty to forty pounds so you can have a lot of them whereas you might have only one crystal puller....

Green: Are we then saying that if a guy wants to make a crystal the first thing he's actually going to consider is zone melting? Stick it in a boat horizontally and—

Draper: The first thing really if someone's going to grow a crystal is probably to ask is what method have I actually got available? . . .

Green: In your lab for instance, though, the kind of thing that everybody would try to use first is zone melting, is it?

Draper: Yes, I think so. In fact all our materials which are interesting have high vapor pressure so in general we use a, sort of, gradient freeze technique rather than zone melting technique. And that's really a consequence of the vapor pressure. It's easier to do a gradient freeze with a high vapor pressure than it is to do zone melting. . . .

Green: We've come round full circle. I started off with the notion that crystal pulling was the method of choice so long as the vapor pressure was less than 1.2 atmospheres but now here you've got a method [referring to gradient freeze] actually, where, presumably going on vapor pressures, actually capable of handling vapor pressures over 1.2 atmospheres.

Draper: A lot over actually—60 or 70 atmospheres—you can grow in that the same sort of thing on which you'd spend half a million pounds for your high pressure crystal pulling. . . .

Green: What intrigues me now is why we've pursued crystal pulling. Didn't we start by—presumably we started because I selected an arbitrary point on this diagram rather than saying—

Draper: Well, no. It is really the best way of growing large crystals providing that there are no problems.

Green: So the first thing you're going to try then is some version of this gradient freeze thing.

Draper: That in general is a very good way to start because you can seal everything up so what's in there is going to be more or less what your crystal will consist of. You know you're not going to have too much problem of things disappearing off and condensing somewhere else[3].... Perhaps I could let on something here and that is that we have a radio frequency generator and a crystal puller but it's very, very rarely used. Because —

Green: You bastard! [laughter] All round the houses on crystal pulling.

Draper: (a) For research purposes you don't generally need very large crystals. (b) All the crystals that we've found that are interesting from the scientific point of view have all got these blasted high vapor pressures. So in fact, you know, we have this white elephant which we always ought to sell but we keep saying, "no don't let's get rid of it because, you know it will come in useful one day".... We're using the radio frequency heating part of it for metallic glasses.

Collins: [pretending to quote from his notebook] "The knowledge engineer gets cross" [laughter].

Green: I've just become a world expert in a lost art of crystal growth. I thought I was going to be able to sell this expertise,... but nobody wants to know about crystal pulling.[4]

Collins: When did you last use it?

Draper: Over two years ago—at least two years ago.

The discussion between Green and Draper concerning the Czochralski technique characterized many of the knowledge elicitation sessions. Green had started by using the chart (figure 10.3) and had decided to work his way through the possibilities contained in the bottom right-hand quarter of it. Draper responded to each query as best he could using book knowledge wherever his practical experience was inadequate and on some occasions apologizing for his lack of knowledge:

Green: [pointing to chart] We haven't got those as far as I'm...

Draper:... They're minor variations. Before we discuss those I'd better find out about something I know nothing about. I know nothing about skull melting and I'm not quite sure what the difference between normal freezing and directional freezing is. Cooled seed is an interesting one—yes.

Green: If you don't know anything about those,... do we need to consider them....

Draper: We've only covered a very small part of the total set of possibilities.

Collins: Wouldn't that part be a sensible place at which to start building the system, and then, you know, there's the possibility of extending it all over the place.

Green: Yeah, I've got some ideas about building it, but I'm still not convinced that the system doesn't say, 'Zone freezing. You can do anything you like so long as it's zone freezing!'

Draper: Why do most watches—mechanical type watches—have jewels that are grown by the Verneuil process, which is another interesting process. It's where you drip molten—where you drip alumina powder with a bit of chromium down through the middle of a gas jet. And there are thousands of these things sort of steaming away in Switzerland producing rubies... Right at the moment I'll pass up the skull melting and cooled seed, because there is something I didn't look up before I came down here.

Collins: Have you ever done either of those?

Draper: No.

Collins: Actually, it mightn't be a bad thing to ask what you've actually done, Bob, out of that list of things.

Draper: Well, we do zone melting, we have done Czochralski in the long distant past. We do gradient freezing a lot. And they're the main techniques we use—Oh, and solution growth we use for a different area.

Collins: But you've never done pedestal pulling or—

Draper: No, because the actual hardware's very expensive....

Collins: Let's take this last one you've just mentioned—vapor-liquid-solid growth. [A crystal is placed in a liquid beneath a vapor atmosphere. The vapor dissolves in the liquid and then grows on the crystal.] Have you seen that happen? How do you know about it?

Draper: I've read about it and I've been told about it by my boss. Now whether it's of any significance, or whether it's just regarded as a curiosity, I don't know.

What suddenly became clear around this time was that in what were supposed to be *knowledge elicitation* sessions Draper was not being used as a human expert but as an intermediary between Green and the already published literature. Draper was being used as a talking book. This, of course, was entirely pointless because Green already had Pamplin's book to work from. The problem arose continually throughout the elicitation sessions, and exchanges of book knowledge took up by far the larger part of these discussions. Even though the futile discussion of crystal pulling had taught us the painful lesson that we should start with Draper's experience, the formal aspects of his knowledge continually resurfaced. Thus, Green and Draper had long discussions of methods of heating (radio frequency heating versus radiant heating); containment methods, including exotic materials for lining boats that had never been used in Bath; the technique of pedestal pulling, which involves using the material itself as the container by melting a pool on the top of a column; the Verneuil method, which is used by the Swiss to grow jewels for watches; and so forth. Based on these discussions, CRYSTAL would simply have been a pale reflection of Pamplin's book. The main reason the textbook was so dominant, I believe, has to do with the conventional hierarchy of knowledge types.[5]

Rules of Thumb

The top of the conventional hierarchy of knowledge is a mathematical understanding of the physical world. Here is an example of the style from Shah (1980, pp. 301–355):

8.2. Theoretical Aspects of Zone Melting

For intelligent applications of zone-melting processes it is absolutely vital to understand the mechanism responsible for the impurity redistribution in processes involving a solid-liquid interface. The key parameters, which need to be defined before any mathematical treatment can be introduced, are the distribution coefficients. These are dealt with in section 8.2.1. For the benefit of new entrants in the field the simple cases of normal-freezing and single-pass zone melting are fully derived together with a review of more complicated cases in sections...

8.2.1. The Distribution Coefficients

8.2.1.1. The Equilibrium Distribution Coefficient

The equilibrium distribution coefficient was first defined by Pfann with the aid of a phase diagram of a binary system with a solute (a soluble impurity) and a solvent (a host material) as components. Figure 8.1a and b [not shown here] schematically represent portions of such diagrams near the melting points of solvents. The equilibrium distribution k_o is defined as the ratio of the concentration of the solute in the solid to that in the liquid C_L when the solid and liquid phases are in equilibrium. i.e.,

$k_o = C_{S/}C_L$

From Fig. 8.1 it can be clearly seen that ... [etc.]

This kind of treatment was of no use in the practical business of growing experimental crystals. To give another example, consider the question of rate of heat loss from the melt. The maximum cooling rate is determined by the need to avoid "constitutional super cooling." This is the solidification of pockets of the separate, pure constituents of the mixture. Pockets of constitutionally super-cooled elements within a crystal will ruin it. Given this upper limit, the rate of cooling affects the speed of solidification of crystalline material. In theory, as the crystal gets bigger its surface area gets larger, so it ought to be desirable to deposit solid material on the surface at an increasing rate. This suggests that the rate of cooling should be increased proportionally to the growth of the surface area of the crystal. After a long discussions with Draper, however, Green, who had hoped to develop a *deep knowledge* theory about the matter (see chapter 9), agreed that this would be too complicated in most circumstances.

Green: So you decide it though, by experience. I mean, are there any rules for—

Draper: There are some formulas, but the problem there is that the odds are that you won't actually have the numbers to put in the formula. It's very nice having a formula that describes it but unless you know all the various parameters, that formula is not—you know, it depends on how many times you're going to do it.

Here Draper is pointing out that a formula may be useful for an industrial system where the same crystal is to be grown over and over again with maximum efficiency, but the effort of measuring and determining the exact parameters for growing a few examples of new crystals is completely out of place in a research laboratory. This can be described in the language of chapter 8 as a case where knowledge from lower down the hierarchy is the most useful.

A similar problem arose out of thinking about the melting points of mixtures. The problem is that if one of the constituents of the crystalline substance has a very high melting point, it may not be possible to find a suitable container material. It may be, however, that the material will dissolve in one of the other constituent elements, making the net melting point somewhat lower than the highest melting point of any single constituent. Effectively this is solution growth using one of the constituents as the solvent. For the system to make use of this, it had to be able to predict the net melting points of unforeseen mixtures in unforeseen ratios. Therefore, we set out to find general rules for combining melting points. This would be an excellent piece of deep knowledge that could be encoded into the system, providing a substantial advance over the knowledge of our expert. As far as we could find out by asking chemist colleagues, however, there are no such rules.

A little thought reveals that the rules certainly cannot be straightforward. Think of the net melting point of equal atomic equivalents of chlorine and sodium; they are both very low melting point constituents yet the net melting point is that of common salt. We were never able to incorporate deep knowledge of this sort even though Green struggled hard to find the relevant relationships. Once more, it was experience rather than deep knowledge that determined the container material. We arrived at the same conclusion after thinking about the maximum pressure that the reactions would generate.

Green: The vapor pressure you get is some kind of linear function of the vapor pressures of the constituents, is it? I can't quite remember now.

Draper: Er, yes, there is that sort of relationship, I think, but what we normally do is look to see which is the highest one and cater for that. So if you are going to grow something that has got mercury or sulfur in, if you're doing a sealed ampule technique, you'd do it in a very narrow one with as thick a wall as possible.

Green: Yep [taking notes], so look at the highest vapor pressure—you don't actually try and establish what is the actual true net vapor pressure of the melt. . . .

And later:

Draper: What you've got to look out for is that the vapor pressure of—the highest vapor pressure of the individual elements before you get to the compound. I think it's the compound can have a lower vapor pressure than some of the individual elements. So you might have, er—

Green: Once you've bonded them together...

Draper: I think that's right so that the, er—typical elements that have high vapor pressures are mercury, sulfur, and phosphorous, and if you are going to grow those, what you've got to look at is the vapor pressure at the highest temperature you think you need to go to with, say, those three elements.... We've got lists of the vapor pressures of all the elements and lists of them at various temperatures.

Green: But it—As a first pass ... what you're interested in is the upper bound on the vapor pressure and that's likely to be the vapor pressure of the individual component—the highest vapor of an individual component at the melt. But the actual vapor pressure might be less even before the crystal's formed because of mixing type things.

Draper: But as a safety thing one would say, "Look at the highest vapor pressure. How high do you think you're going to go in temperature? What's the, what is that pressure likely to be?" and then from that make some kind of judgment about how thick a walled ampule you might need to use.

Collins: That's the rule of thumb, is it? You always go for the highest conceivable vapor pressure rather than—

Draper: Yes.

Problems of Knowledge Elicitation

These, then, are the difficulties of knowledge elicitation. There is too much respect for theorized, textbook knowledge, and this made it hard for Green and Draper to talk of practicalities. *Talk* about procedures is far greater in volume and content than program rules or text and far more revealing of doubts, qualifications, and uncertainty.[6] Talk about crystal growing is full of uncertainties and "uhms and ahs." With the abstracted certainties of the textbook and the idea of deep, scientific knowledge all the uhms and ahs have long disappeared. The attractiveness of textbooks, expert systems, and schoolroom science in general is that all doubts and the associated responsibilities of decision making in conditions of uncertainty have been resolved. This is the attractive part of both bureaucracy and the tyranny of machines.

Another problem that is well known to anthropologists and social scientists with an interpretative frame of mind is the invisibility of that which is too familiar to us. There is no reason to expect an expert in crystal growing, or any other esoteric area, to have phenomenological self-consciousness in respect of his, or her, everyday life. As anthropologists and philosophers know, developing self-consciousness is a difficult and painful experience. It may be that textbook knowledge is more easy to speak about precisely because its abstract nature lends it a perpetual and inescapable unfamiliarity. It is always outside us and therefore always a visible feature of our mental lives. The hierarchy of knowledge values the theoretical above everyday understanding, and what is more its very unfamiliarity gives it salience.

Some very simple rules for knowledge engineers follow from this account. First, the knowledge engineer should attempt to gain enough experience so as to be able to recognize when textbook knowledge rather than experiential knowledge is being offered. In our knowledge elicitation sessions it was usually I who had to cut short the recounting of the text, but I was in a position to do this only because I was spending time in the laboratory. Second, for the purposes of knowledge elicitation, the expert should be questioned by a potential user of the system. Only someone with the skills, or lack of skills, of a potential user will naturally ask questions that are pitched so as to render experts' worlds strange to them to just the right degree. There is no way in which the whole of an expert's knowledge can be elicited, but there is also no systematic way in which a knowledge engineer could extract just the right amount. The expert will tend to volunteer less than a novice will require because of the invisibility of what is needed by the novice but overfamiliar to the expert. Knowledge elicitation and prototyping of part-completed systems should take advantage of the different saliences within the worlds of novices and experts by arranging the right kinds of interactions between them. The knowledge engineer needs two volunteer helpers: an expert and a user. It is not a matter of extracting all the knowledge—an absurd project—but a matter of eliciting one description of the world from among the many that are available. It is a matter of selecting the description that has the best chance of suiting the user.

The Expansion of Knowledge at the Workfront

As Draper was pressed to move away from the textbook account into the world of practicalities, description began to ramify.

Draper: What we haven't really talked about is feeding the input on apparatus into the system. You know, what are the different sort of container materials available. ... If you had a zone refiner you might contain the thing in a Pyrex boat, a "Vycor" boat, a molybdenum boat, a graphite boat, pyrolytic carbon or pyrolytic boron nitride boat. Ah, the actual—

Green: Ah, so in other words, then, within these main categories there are lots and lots of subdivisions in terms of the fine tuning of these things.

Draper: That's right, you know: What are the apparatus parameters? How are you going to heat it up? Are you going to heat it up by, say, a normal resistive furnace and, if so, what sort of elements are you going to use? Is it going to be something like a "Kanthal" wire or does it need to be a platinum winding, or do you need to go to radio frequency heating? How do you actually contain the system if it's, say, a zone-refining method? Do you use Pyrex tubes or do you have to use silica tubes which have different melting points, or softening points? Is it necessary to see what is going on? ... What mechanical transport is needed if it's something where something is moving? If so, does it matter how fast you go? How do you control the temperature gradient and what sort of temperature gradients are needed? That's another sort of point. Ah, if something is heating or cooling does it matter how fast or how slow it's cooled and does it matter whether it's linear or nonlinear?

There's an awful lot to say [laughs]. When you start listing it out there's a hell of a lot in a relatively narrow field. You know, "What sort of atmosphere? Do you need a vacuum, a reducing gas, an oxydizing atmosphere, an inert atmosphere?" ...

This last comment of Draper's expresses quite succinctly the way that the field of crystal growing unfolds again and again as one tries to articulate everything. It unfolds like silk scarves from a conjurer's hat, but more so because each new scarf contains as many scarves as the hat itself. Green's struggle was to reduce this cornucopia of particulars into some general rules. It turned out that for the system to be useful the overarching schematization of the textbook was not the appropriate starting point—rather, we needed a more detailed description of the narrow areas of expertise available in the Bath laboratory. But even these narrow areas threatened to become unmanageable as the interviews progressed. One might think of Green's attempts to find deep knowledge that would determine net melting points and net vapor pressures as an attempt to replace the chokingly rich particulars of the characteristics of individual crystals with some simple overarching rules.

The complexity we have encountered so far is, as previous chapters have led us to expect, merely the beginning. The silk scarves did not really begin to pour from the hat until I entered the laboratory.

11

Apprenticeship—First Steps in Crystal Growing

My hands-on learning began in April, 1984. To start with I visited the physics laboratory on seven occasions from April 5 to April 17. My guess would be that the image that comes into the mind of readers when they think of growing semiconductor crystals is a hard-edged, clean, technocratic laboratory. Well-ironed white coats and shiny machinery is certainly the first image that would come to my mind—a sort of scaled-down version of the Silicon Valley microchip factories that we see on television. The product ought to be angular, sharp, and shiny; the very word *crystal* is a metaphor for clarity and purity.

As in the all of science and technology the crystalline image is easy to maintain only when we are distanced from the muddle and anxiety of the laboratory. For scientists, a few words in a journal article can represent months or years of effort. These words are a neat little parcel that, it seems, anyone can pick up and carry away. We forget the confusion involved in reversing entropy—in taking a fog of novelty and molding it into bounded packages of similarity and difference.

For the crystal-growing apprentice the first step is disenchantment.[1] The product of a growth-run for a new crystal does not look like something from a Silicon Valley clean room; it looks like a small piece of dirty gravel. These things are carefully stored in labeled glass vials because if they were dropped on the floor they would never be identified again amidst the rest of the dirt. Distance lends enchantment in crystal growing as in everything. Semiconductor crystal growing is quite unenchanting after a couple of days doing it.

The demystification that comes with the first sight of the product is a crucial part of apprenticeship because it begins to provide a criterion of what counts as success. To make a thing that might have been prized from the tread of a shoe is much less daunting than making a *crystal*. As I will try to point out throughout this chapter a vital part of learning a skill is learning what is to be achieved.[2] This is not always a matter of learning how easy things are; to master some skills it is necessary to know that they are difficult. For example, I cannot imagine many people learning to ride bicycles unless they had seen one ridden. If they did not know what a bicycle was they might eventually try to balance on it, but unless they knew it was a difficult thing to learn to ride—something that takes at least an afternoon of continual effort—they would soon eliminate transport as its function. After an hour of bruising and painful disappointment only a fool or a genius would not conclude that if a bicycle were meant to stand upright there would be a third wheel. Likewise I would not have learned to seal a quartz ampule without being shown that it could be done and therefore deducing, from my own initial failures, that it was a hard thing to learn. On the other hand, I would not have dared to make a thermocouple had I not been shown that it was very easy.

Bismuth and Arsenic: Rules and Practices

On the first of my visits to the physics laboratory I spent several hours watching as Draper prepared to grow an experimental crystal of bismuth with half-percent arsenic by weight. I had intended to help Draper grow the crystal as the first stage in my apprenticeship, but I immediately discovered that I did not know enough even to be able to start. For example, we needed the raw chemicals to begin with, but I did not know where to find them. The laboratory was a completely unfamiliar environment to me—nothing was ready at hand. A cook who tries to work in an unfamiliar kitchen knows that it is difficult because the interlocked movements and coordinated timing required for efficient cooking cannot be managed when the salt, flour, sieve, or cleaver is not there when the hand reaches for it. Part of the skill of cooking is in reaching for vessels and ingredients and using them smoothly and continuously as an integral part of the cooking itself. The physics lab was like an unfamiliar kitchen. I needed Draper just to find my way about.

What about the proper method for growing bismuth-arsenic crystals? This seems to be less of a local matter. CRYSTAL was designed to give universally applicable advice on crystal-growing methods. What does CRYSTAL say on the subject? Well, the high vapor pressure of arsenic rules out the open-boat method, but the propensity of bismuth to expand on freezing rules out the ampule method. Thus CRYSTAL resorts to its catch-all solution:

==> (TRY PARTIAL-COMPOUNDING OR TRY SOLUTION-GROWTH, THE METHOD AND CONTAINER WILL DEPEND ON THE PARTIAL COMPOUNDS OR ON THE SOLVENT)

Partial compounding, which will be explained in chapter 12, will not work in this case because it is an ampule method. This leaves some sort of unspecified solution growth as the only answer. What is authoritatively clear is that neither open boat nor ampule will do. What method did Draper decide to use? He used an open boat![3] This is what Draper said to his apprentice on the matter:

In this case we couldn't grow it with a Czochralski technique because one of the problems you have with the Czochralski technique—that's the crystal-pulling technique—is that you need to be able to see what is going on. Now, with the arsenic when you heat up a melt with a volatile component, which is the arsenic, the arsenic—the book may omit to tell you this—the arsenic comes off and immediately coats over all the cool surfaces and you can't see what's going on.

Now, for materials which contain arsenic which have high melting points, such as gallium arsenide, what you can do is to effectively cover the molten material with a transparent glassy layer. Normally the material used is boric oxide. It's transparent—you can see through it—and normally what you do is put an over-pressure on it and that stops the arsenic coming off. And you can sort of crystal-grow through that. . . . You can grow the crystal through the "encapsulant"—the over-pressure keeps the arsenic in place—and then afterwards the advantage of the boric oxide is that you can dissolve it in water. Now the problem with encapsulants is that there's really only one, and that's boric oxide, . . . and boric oxide doesn't really become liquid until about 800°C. Below that it's a bit like toffee . . . so in this case, because bismuth is a low melting point material we can't actually use it . . . so we're stuck.

It is worth noting at this point that Draper's explanation to me, the novice, of why we don't use the Czochralski method is technical and universal. Only much later did I learn that the Czochralski method had not been used at Bath for at least two years and that it was probably never under serious consideration.

He went on:

So what we'll have to do is to use a technique . . . If we have a small system we can probably tolerate some loss of arsenic, and the best thing to do is to use a horizontal boat.

Now the problem with the bismuth is that it expands on freezing. The easiest way, perhaps, to grow something might have been to use a vertical . . . technique where you seal up the material in an ampule and then you lower it down in a furnace and directionally freeze it. The only problem with that is that since the stuff expands on freezing you'll probably break the ampule so what you need to do in that case is to use a horizontal technique where you have a semicircular boat which doesn't constrain the material when it actually freezes. If one does it in a horizontal system, which is, you know, not too large, you can tolerate the arsenic loss. . . .

In this case it's an open system; we won't be sealing it so what we're going to be looking at—one of the things we want to know—is can we actually grow—can we see at the end of the day whether we've got a single crystal or, if not, how big the grains are and whether they are big enough to actually cut a crystal out.

To reconstruct, this reasoning is a pragmatic compromise unanticipated by CRYSTAL. Rather than do anything too sophisticated and troublesome, such as the solution growth that CRYSTAL would suggest, Draper decides to use the open boat on the assumption that the losses of the volatile arsenic will not be too great so long as the whole system is fairly small. The boat method has the great advantage for a brand new experimental crystal that the ingot can be easily inspected; we will be able to see how the process is going. But this was not all there was to Draper's method. He decided to try to minimize arsenic loss in another way:

Draper: The other thing is that, what I'll do is melt this down in a small volume first. Once the arsenic is dissolved in the actual bismuth, when you put it in an open boat there's much less chance of it coming out. So rather than just putting the individual elements in an open boat, what I'm going to do is put it all into a small silica ampule and then dissolve it up.

Thus the method adopted was a hybrid that broke all of CRYSTAL's rules. The materials would be melted up in an ampule to intermingle them as much as possible—and we would not worry about whether the ampule cracked—and only then would they be dumped in the boat. No such possibility was ever mentioned in the knowledge elicitation sessions nor was such a possibility encoded into CRYSTAL.[4]

The idea of mixing up the compounds a bit to minimize vapor loss is, of course, taken from well down the hierarchy of knowledge types. That Draper's method of choice broke both of CRYSTAL's main methodological precepts is interesting. Remember that CRYSTAL was built by reconstructing the logic of crystal growing from talk—

by imposing a systematic structure on the field. The fundamental framework for CRYSTAL was hacked out of the choking details by Green. Even though the knowledge elicitation conversations, as represented in the last chapter, were at variance with what we were *concurrently* doing in the laboratory, this did not come to affect the process of abstraction of the rules for CRYSTAL's program. It could not have done without wrecking the structure that Green had wrested from Draper's reluctantly recollected but still hugely incomplete accounts of practice within the Bath laboratory. Green had discovered a workable resolution to the tension between the rule base becoming an endless list of special cases and the need for a structure that could be encompassed within the span of a small project.[5] But what should there be in a system designed to replace an apprentice master? Is it conceivable that everything in this chapter and the next should be included? And would that be enough?

Weighing Out: Approximation and Manual Skills

Given the method that Draper had chosen, the first thing we had to do was weigh out the correct quantities of bismuth and arsenic and load them into an ampule. Bismuth is a metal and metals can be had in a variety of forms such as powder, strip, shot (which looks like shotgun pellets), or lumps. We used bismuth shot. Draper found the bismuth in the relevant cupboard—itself a task of half a day or longer for the untutored. The next question was: "How much should we use given that we wanted a crystal of unspecified size?" To solve this problem Draper referred back to some of his own handwritten notes covering previous trials involving bismuth. From these he discovered that about 30 grams would fit into the sort of boat that was to be used for this run. Thus the size of the boat—the apparatus to hand—had decided the volume of the crystal and previous experience decided the weight of bismuth that would fit the volume. None of these solutions could have been fitted into even the largest expert system. The user would have been expected to supply the answers—something that was easy for Draper but impossible for me.

The arsenic—metallic arsenic strip—was found. Arsenic is poisonous. How was it to be handled? It flashed across my mind that we should use masks, gloves, breathing apparatus, or remote control. Draper explained, however, that metallic arsenic, being relatively insoluble, was not as poisonous as arsenic salts. Nevertheless, "care must be taken!" White coats and tweezers were the order of the day and we had to wash our hands thoroughly before eating.

Weighing was the first operation I watched. We used a sensitive electrically powered chemical balance capable of weighing grams to four decimal places.[6] The chemical balance had a series of different colored knobs and dials used to switch it between ranges of sensitivity. Each range was an order of magnitude different to the next (that is, it represented the next decimal place in the fraction of a gram).

Chemical balances have a rest position where the balance pan is supported by a stand while the actual weighing is done by raising the pan from the stand. If the pan is significantly out of balance it will oscillate when raised and the oscillations can go on for an irritatingly long time. A purely manual skill that had to be mastered early on was damping these oscillations. This could be done by lowering the pan to near rest position and then raising it very gently.

To weigh out a fixed quantity one places an amount in the pan, weighs it roughly, and then adds or subtracts a smaller quantity, reiterating the process until the desired mass is achieved. There is a philosophical nicety involved in weighing out to a specified target. Unless one can judge the weights by eye one always adds on (or subtracts) too much or too little. This is a practical version of Zeno's paradoxes. In principle one may weigh for ever without getting it right. The inescapable and really rather striking logic is that if you can't judge the weight of something by eye, then you cannot use a chemical balance to weigh out a specified quantity. Even to reduce the number of iterations required to weigh out with *reasonable* accuracy requires enough manual and perceptual skill to judge a mass well enough to get close to the right amount and then physically to separate that amount from the native form in which it comes (for example, wire strip). None if this is easy.

Once the skill of judging and cutting has been mastered the question remains: "Just what accuracy is required?" Chemical substances are effectively indefinitely divisible. The design of the chemical balance sets a limit to the resolution with which the substance can be measured (it digitizes the world, as one might say), but one cannot physically break the world into pieces that small. At the fourth decimal place it becomes very difficult to separate the tiny particles from the tweezers. As with the use of a calculator, judgment about appropriate levels of inaccuracy is unavoidable.

This is how Draper and I solved the matter of how accurately to measure out the chemicals:

Draper: It's .93, and then the next figure is there which is .931 and the figure after that is going to be an 09 probably, or possibly a 1. You're getting on to the levels of inaccuracy of the machine here, really. It's either .9311 or .93109.

Collins: Are you interested in the [fourth] decimal place?

Draper: Uhm, not really. I mean, you aim for it and then hope that the one before that is at least right. Uhm, you get to the stage where the accuracy with which you can chip lumps off, or whatever, is difficult. . . . It's really a question of percentage. The smaller the quantity of stuff you're actually weighing out, the more important that last figure is. If you're weighing out 100 grams or 1,000 grams it really doesn't make much difference.

At this point we take a second measurement *without* altering the content of the pan and discover that the reading differs by 1 at the *third* decimal place. The real accuracy of the balance is substantially less than its theoretical accuracy. The scaling knobs are as misleading as the maximum speed on the speedometers of most cars. There is, then, not a lot of point in agonizing about the fourth decimal place after all. Draper is not too worried, however:

Draper: If you're weighing ten, twenty grams it [this machine] is adequate. If you're—it's not as—you know, if you're weighing less than a gram it's not really that good, but if you're weighing more than a gram—

With the bismuth-arsenic crystal we were weighing much more than a gram and there was no need for great accuracy anyway.

To do the weighing we used tweezers to handle the arsenic. We put the materials in flexible plastic cups to be placed on the pan scales. (The cups were weighed beforehand so that their contribution could be subtracted.) On matters of purity and accuracy Draper commented:.

Draper: If it was a very high purity that we actually required, then all the implements we use would probably be specific to a certain element. We would only use, you know, a pair of tweezers on a particular element, we would not use tweezers on different things or different elements. . . . In general we are not concerned with that degree of accuracy. . . .

The actual accuracy is not too important with this because we will get some arsenic loss. What we can do at the end of the day is actually determine the amount of arsenic that's actually left in the material rather than what we put in in the first place. . . . [weighs sample]. That's accurate to one part in 3,000 odd—plenty accurate enough for our particular use this time. . . .

Now, there are various losses that occur on the way, so there will be some of the arsenic that evaporates and, uhm, forms a film on the ampule. . . . One of the problems here is, you can lose bits of the elements on various parts of the course. It's rather like the Grand National; you lose various bits on various parts of the course. When you actually pour the stuff into the ampule some will stick to the sides of it. When you heat it up you will get a thin film formed. So all you can say is how much you tried to put in the first place and then at the end of the day you can perhaps use some analytical techniques and say what you actually ended up with getting. . . . So although we can put half-percent by weight in now we don't necessarily know that we will get half-percent in the final crystal. But, you know, for a first attempt it will be good enough. But we might find that we have to put in more than half-percent of the arsenic to get half-percent in the actual final material.

It would be difficult to put this kind of run-specific reasoning into an expert system. Draper is doing an *act* of crystal growing that could only be specified in terms of rules given hindsight. The exact way the materials were to be weighed and handled depended on what was to hand, who it was to be grown for, and the fact that it was a first experimental run. The conditions were unique and impossible to anticipate. In Suchman's (1987) terms, it is the situatedness of this act that is its salient feature.

Filling, Mixing, and Melting

Draper now set about making the ampule in which the materials would be initially melted and mixed. For low melting point substances such as these, CRYSTAL would prescribe easily handled, cheap Pyrex glass. Draper decided, however, to make the ampule out of quartz [silica glass]. I asked him why:

Draper: We want to heat it hotter than the actual melting points to make sure that everything is really thoroughly and utterly mixed in. The other thing is that silica doesn't shatter. Pyrex, unless it's carefully annealed, is quite easy to actually break due to thermal stress whereas silica you can ill-treat.

The point is that because we were using the ampule for mixing rather than growing there was no need to heat and cool it gradually. To save time Draper wanted to heat and cool the ampule quickly and for this, quartz was the best. Once more, local circumstances were not anticipated in CRYSTAL.

The next perceptual skill I learned was picking a piece of silica glass from the rack of assorted glass tubes by reference to its color, which is white rather than greenish yellow. Making a quartz ampule is not easy. Later I was to try it and fail. There are a number of important points. Silica fuses at a high temperature and gets extremely hot as it is worked. The first operation, then, is to fuse a piece of scrap quartz to the end of the working piece so that it can be used as a cool handle. The ampule itself must be made extra thick and strong at the bottom where it curves round. There will be a section of the quartz that tends to crystallize as it cools (visible as a bright white area) and this needs to be remelted (annealed) or it will be weak. Finally, the ampule will be narrowed (necked) a little at its open end so that the final task of melting it closed after filling and evacuating is less difficult.

After the ampule was formed Draper cooled it with a blast of compressed air played around the outside. Compressed air is available on tap in the physics laboratory. He pointed out to me that he was careful not to direct the compressed air into the ampule because, having passed through a pump, the air was marginally contaminated with oil. Should this be a rule in CRYSTAL?

Now we load the weighed elements into the necked ampule. Draper places a large sheet of fresh polyethylene-covered paper beneath the ampule to catch anything that misses the neck. He takes a little flexible plastic disposable container (also called a boat) that had been used to weigh the arsenic. He squeezes the container so that it forms a point at the end—effectively a kind of funnel-shaped guide—then holding this in his left hand with the point directly over the orifice of the ampule, he taps his left hand with his right hand so as to gently vibrate the boat. In this way the contents of the boat—arsenic strip—are persuaded to drop neatly into the ampule without loss.

The operation is now repeated with the bismuth shot. This tends to jump out of the boat and bounce out of the neck of the ampule. Draper explains that it's no use using a funnel as the stuff invariably jams in the conical section. I ask permission to load the bismuth myself. This is my first practical experience in the art of crystal growing—putting bismuth shot into an ampule. There is no textbook that tells one how to do this.

Draper makes sarcastic comments as I do my best but the pieces of bismuth shot continually miss the ampule and fall noisily onto the paper placed beneath. (The sound is nicely captured on my tape recorder.)

Draper: One in, three out, so far. Ten percent. Up to forty percent [and so forth].... [and then, sarcastically quoting into the tape recorder:] "Harry is now scraping together the results of our combined first attempts and we'll see whether the percentage gets better as he perfects the art.... One thing he didn't notice when he was commenting on how I was doing it was that I actually had one finger touching the top of the tube so that when you actually vibrated the hand just by tapping it you didn't actually vibrate it very much and that there was a sort of relationship between the top of the ampule and the boat and this was because of a finger that he hadn't noticed, actually." That's right. "Now he's adopted that style and the results are much better—ten percent loss rather than ten percent going in. Just a little thing like that makes all—"

Here then, we have another insight into the science and art of crystal growing. I have partially described what is involved in putting the substances into the ampule. Look at the length of this description—and it is a description of a part of crystal growing that does not appear anywhere in textbooks or in conceivable expert systems. Resting one's finger on the ampule as you vibrate one hand with the other is the crucial technique! Why is it nowhere mentioned? After all, the fact is that if you cannot do it, you cannot grow crystals.

The next thing to be done was to evacuate and seal the ampule. Evacuation involved three separate processes. First, the ampule was filled with hydrogen several times and quickly and partially evacuated with an ordinary mechanical vacuum pump. In effect, this amounted to rinsing out the ampule with hydrogen—a reducing agent—so as to remove as much oxygen as possible. The idea was that any residual gas left after the final evacuation would be hydrogen rather then oxygen. Then, the ampule was connected to a diffusion pump for thorough evacuation. The diffusion pump is a complex array of glassware with oil bubbling and dripping through it. It behaved in terrifying ways; I never did work out what was going on. I would not be able to use a diffusion pump without considerable extra instruction. Third, Draper explained, the diffusion pump was not effective at removing water vapor, so he arranged that the outlet pipe from the ampule would pass through a cold trap. A cold trap is a section of pipe that is dipped into a vacuum flask filled with liquid nitrogen.

The nitrogen was stored in a large vat in the corridor. The vat was on a stand with gimbals so it could be tilted and poured. Pouring had to be done gradually so as to slowly cool the receptacle. To liquid nitrogen, a room-temperature flask is like red hot iron to water. Any suddenness would result in a boiling spray of deadly cold liquid nitrogen that would produce a painful effect on the skin—or a nasty "burn" if it were to remain cupped in the palm or held in a fold of flesh.

We set up the cold trap and the diffusion pump.

Draper: Right, now at this point what we must do is just wait. We have to wait about half an hour or so for it to pump out.

Collins: You say we've got to wait half an hour. How do you know it's half an hour?

Draper: From experience, really. It's a very small tube that you're pumping out. But then again the actual diameter is very small so it does take a fair bit of time to actually come out through that small tube. The thing is, we have flushed it out several times with hydrogen, so hopefully what's left in there is now hydrogen rather than oxygen.

Now we come to the matter of sealing the ampule. The importance of one, seemingly inconsequential, decision becomes clear. To seal the ampule we must use a blowtorch on the quartz while it is filled with volatile and poisonous chemicals. The fact the arsenic was put into the ampule first means that it is at the bottom so that the most volatile and most poisonous substance is farthest from the flame. The ampule is long enough to allow the point of application of heat to be a couple of inches above the top of the bismuth. The contents had been firmly tamped down before evacuation—another small but vital matter of order and timing. A further precaution is to wrap the top of the ampule, between the point of heating and the chemical contents, with a piece of Kaowool (an inert mineral cotton wool) that had been soaked in cold water.

The ampule was to be sealed with a blowtorch. To use the blowtorch one needed to know about which nozzle to fit, which gas cylinders it should be connected to, how the cylinders should be turned on, and in which order, how to control the flame of the torch, and what size and character of flame was suitable for working with quartz. Then, there was the problem of holding the ampule near to the immensely hot flame—wearing the right goggles and using the correct protective glove—and the matter of speed of heating and dexterous manipulation of the objects.

Draper: The problem here is to try and spread out the area which is actually . . . I don't want an inward bursting bubble which is possible. OK, I've started rotating it which will help to seal it off.

Collins: I see, an inward bursting bubble happens if the silica becomes too thin and the vacuum pulls it in.

Draper: That's right.

The idea is to get a nice even thickness of glass (or silica) so that it won't burst inward under the initial vacuum or outward when the pressure of the evaporating arsenic develops.

Now that the materials were sealed in the ampule one would normally use a furnace to heat it, but because we were only trying to mix the ingredients Draper decided to heat it with a gas torch to save time. He explained that this would be fairly safe because the vapor pressure of the arsenic would only be little more than one atmosphere. With some elements in use at Bath the vapor pressure might be of the order of 10, 20, or 30 atmospheres so a furnace in a fume cupboard would have to be used for safety. The contents of the ampule became liquid without an explosion of arsenic vapor and Draper agitated it so that everything was well mixed. As it cooled he held the ampule at a slight angle to the horizontal so that the liquid was spread along the tube.

Draper: I want it to be compact, but I don't want it completely filling one end of the tube because it's likely to crack it when it freezes.

The mixture cooled without cracking the ampule—again, something that should not happen according to CRYSTAL. The rule that "mixtures that expand on freezing crack ampules" proves inadequate under these particular circumstances.

We look at what we have inside the tube; the effect is not impressive. The mixture looks quite scummy on the surface. We conclude that this is oxide film; there was certainly some oxide on the bismuth shot. Some arsenic has been lost in the form of a condensed film on the inside of the ampule. Draper says that the quantity of arsenic that has been lost is not great because it has been heated from the top and therefore the coolest part was the melt. He hopes, therefore, that there will not have been too much condensation. The location of the heating was again something I had not noticed but which now appears to be of some consequence; Draper had heated the ampule above the substances and this had melted the bismuth before the arsenic so that as the arsenic volatilized it was more likely to become mixed with the bismuth than to recondense on the ampule. Another rule missed!

The next task is to open the ampule and extract the contents. There is some danger of an implosion when the evacuated ampule is opened. First, Draper scores the ampule with a diamond wheel. He wears a face mask for this operation. Then he wraps the ampule in a paper tissue before snapping it by hand. The cooled ingot comes away from the glass cleanly. We had originally planned to etch the ingot at this stage so as to remove oxide film. Now, however, it is not so clear that this is necessary.

Draper: Etching has plus points and minus points. If you etch it, you've wetted it and it takes quite a long time to dry the sample out. So in fact, with this, if we can get this out in one lump we just might physically wipe it just to remove any loose oxides that might have formed.

He wipes the surface of the ingot with a tissue, pointing out that a lot of the "crud" that was on the ingot has remained in the ampule. We now seem a long way from the careful measurements we made with the chemical balance. What has happened to our detailed considerations of purity?

Now we are we ready to move on to growing the crystal.

12

Growing Crystals

Bismuth-Arsenic Using Zone Melting

In zone melting the mixture is put into a *boat* that is semicircular in cross section. In Bath, boats are about a centimeter across and about six or seven centimeters long with rounded ends. Of materials from which boats can be made there is much to be said.[1] In this case, however, Draper had already decided what to use:

Draper: If the material wets the boat and sticks to it, you have to break the boat to get the material out and that obviously strains the crystal and destroys the boat. . . .

Collins: Do you know what sticks to what—what wets what?

Draper: What one happens to know is that very, very few things stick to pyrolytic carbon so, if you can use pyrolytic carbon so much the better.... [You cannot use it is if there is danger of oxidation because it will burn.]

Collins: So do you use it on all other occasions?

Draper: Well, the major disadvantage of pyrolytic carbon is the cost—it's very, very expensive. So you try not to break it. Pyrex is probably the cheapest but it's very difficult to make the boats out of it. This one, it's already made, so you don't have to worry about it.

Thus, in practice, the choice of boat material was pragmatic. Wetting and sticking is a problem much discussed in the textbooks but it is hard to know if some new mixture will wet glass. It seems not to have wetted the silica glass ampule so it probably would not wet a glass boat. There is a pyrolytic carbon boat ready to hand, however, so "Why worry?" The theoretical knowledge of materials discussed in the knowledge elicitation sessions is referred to hardly at all. CRYSTAL would advise a boat made from Pyrex.

Only now can we move to the zone refiner (or zone melter). To see the apparatus was, once more, to be demystified. *Zone refining* is a big, clean, impressive technique used to purify silicon crystals in the semiconductor industry. I remember reading about it when transistors were first invented. The apparatus at Bath, however, is a small and simple homemade device.

The material to be processed is placed in the boat. The boat is placed inside a horizontal glass tube, perhaps 30 centimeters long and 4 centimeters in diameter. The boat has to be pushed from one end until it is roughly half-way along this outer tube. The glass tube is surrounded at one point by an annular radiant heater (wound with coils like an electric heater element). The heater is only about 1 centimeter in width and its heat is narrowly focused. The heater is connected to a screw mechanism with an electric motor and timer and, once set in motion, it travels slowly along the length of the boat, melting a narrow zone as it goes. The mixture in the boat is solid before the heat reaches it, then it melts, then solidifies again as the heater element passes on. The heater will move from one end of the boat to the other in about twelve hours. Impurities tend to migrate to the molten zone.

In zone refining, impurities are encouraged to travel along with the molten zone. After a number of passes in the same direction most of the impurities will be concentrated at one end of the ingot; this end can then be discarded. For growing crystals, the "impurities" are the important thing. In this case the half-percent arsenic needs to be evenly distributed through the ingot. It is the arsenic *dopant* that will give the crystal its desired electrical properties. Even distribution is achieved by moving the heater backwards and forwards rather than making repeated passes from the same end.

Collins: How many passes do you make?

Draper: Enough until we have a sample that's big enough to use.

Collins: And you have to just keep looking at it, do you?

Draper: Yes.

Collins: Presumably, you can take it out and look at it now and again.

Draper: Yes. Luckily, with some materials you can actually see from the surface how many different orientations [crystals] you have got.

Ideally one would prefer that all, or nearly all, the material develops into one single crystal exhibiting only one orientation. In the easy materials the crystals are visible in the same way as on the surface of zinc [galvanized] plate. If there are too many small crystals, the ingot can be replaced for another pass of the heating element. Each pass takes about a day.

Because the direction of the passes is important, it is equally important to keep track of the spatial orientation of the ingot when it is taken out for examination. If it is put back the wrong way round the next pass of the heating element will tend to purify rather than distribute the dopant through the mixture. When an inspection is to take place some peculiarity or flaw in the ingot must be noted before it is completely removed from the tube so that it can be replaced correctly. This again is something that a beginner might expect to get wrong at least once or twice but that would never be mentioned in books or expert systems.

The Zone Melter

The zone melter is flushed out with domestic gas—mostly hydrogen. The tube of the zone melter is quite dirty, but Draper says that it has been used for growing arsenic compounds previously, so any impurities will not be damaging. We have now lurched away still further from the exactitude that preoccupied us at the weighing stage and from the overarching concern with purity that will be found in the texts.

The box that contains the zone melter is kept warm through the night by a thermostatically controlled electric lamp. There are some endearing safety features. For example, to safeguard against flooding a soluble aspirin tablet is used to hold closed the microswitch that controls the power; water will dissolve the tablet and open the switch. The apparatus is in a plywood box; it is grubby and untidy. It is a long way from Silicon Valley. Once more it is the *tolerances* that an unskilled person like myself cannot grasp. I don't know what a proper zone melter ought to look like. I would have spent days trying to clean the apparatus before starting.

The next difficulty is setting the current that passes through the heater. It must provide enough heat to melt a zone, but the zone should be as small as possible. Draper sets the current to 7.8 amperes. I ask him how he knows the setting. He says it is experience of this particular apparatus—and of course it varies depending on the mixture—its melting point and its thermal conductivity. Setting the right current would be at least a full day's operation first time through. One would have to turn on the heater, wait for the compound to melt (perhaps fifteen minutes), check to see if the zone is the right width, alter the current, wait again for the system to reach equilibrium, and so forth. Again, experience and judgment are of the first importance; the paradox of weighing out is encountered once more in the problem of setting a heater.

The next consideration is the speed of traverse of the heater.

Draper: In order to avoid this constitutional super-cooling problem we go fairly slowly. That guarantees that you avoid it. You could do a lot of experiments to find out how fast you could go to avoid it but it's not worth the hassle. If you just do it slowly, you know you'll avoid it.

We now set the zone melter in motion. The normal procedure would be to allow perhaps half a dozen passes before examining the crystal. The crystal can be examined either by peering through from the outside or it might be taken out of the boat and looked at more carefully. Etching the surface might help:

Draper: If you etch it then the different crystallographic orientations etch up in different ways and reflect light to a different extent so that you can actually see where all the different crystals are. And you can then see without actually cutting it up how big the grains are—you know whether they're big enough to cut the sample out you want.

After four passes we return to examine the ingot. We peer at it through the dirty glass tube of the refiner.

Collins: I can see something that looks like fairly large crystals in this now—long ones. Do you think those are crystals? Do you see what I mean? There's some that—it's as though they divide the thing almost in thirds widthwise and they stretch for quite a long length of the ingot.

Draper: Yes, I mean certainly on the, eh, nearside if you like, it looks like more or less half the width of the ingot and over about two-thirds of the length there's a single crystal. The part on the other side does look as though it's got some small crystalites within it. There are some smaller little grains going in this direction. You can see if you get the light right.[2]

Collins: I think I can see them, yeah. Yeah —yup.

Draper: Otherwise the surface looks very smooth. The cracks in the surface skin do look like cracks in the oxide without being actual cracks in the actual material.

We left the material to undergo a few more passes. As pass followed pass, however, the surface appearance of the ingot became more and more scummy and opaque.

Collins: There's so much crappy scum on the surface that it's hard to see if there are any crystals or not. Take it out and etch it?

Draper: Take it out and etch it and see how we're getting on.

We take the boat out of the refiner, using a cunningly shaped piece of wire, and look at it in the light.

Collins: Does that look promising?

Draper: It's, uhm, difficult to—just holding it up to the light by the window you get, ah—in directional light, uhm . . . The material looks quite promising for cutting up . . . The bit that you can see this end through the oxide [scummy film] looks quite smooth.

If we decide to put the ingot back for further runs, another decision will have to be made. Should we turn the ingot upside down? The bottom surface of the ingot is indented with the impressions of gas bubbles and these could be removed by remelting it the other way up. On the other hand, this would put the surface scum on the bottom and as this rises to the surface it would provide new nucleation sites for lots of small crystals. Again, it is a question of judgment based on experience, not science.

We decide to etch the ingot. Etching is again a matter of skill, experience, and trial and error. We try a series of etching agents of increasing strength until we get one that works.[3] Some of these agents need very careful handling. We succeed in ridding the ingot of its surface scum by etching it with concentrated nitric acid. We examine it again.

Collins: Ah! What would you say about that, then?

Draper: Yes, quite decent-sized lumps there. . . . Looking at this now, there's a large grain there, quite a large grain here. . . . Let's hold it up to the light. Yes, I mean all that far end is one. [inaudible] Yes, there's certainly about, yes. There's another grain there and there. But because it was so cruddy I think one might, in fact, put it in again.

Collins: Is that one there?

Draper: Yes, yeah. . . . Well, I think that we can do even better having etched it. [Then, after a few more minutes close inspection] Yeah, this does look very promising. It might be worth cutting that up. . . . Er—yes, the end of that could very well be—that whole lump could very well be a whole single crystal. So we might get that cut in half and have a look at that. . . . We've definitely made a lot of progress. . . . I don't know why the top is not smoother. . . . What it is, it makes it difficult to ascertain whether there are other crystals in there because the surface isn't terribly smooth.

In the end, Draper decides to cut the ingot up and look at its cross section.

Draper: Cutting it into sections is a sort of well-known practice for seeing where the grains go inside. If you look at the surface and look at all the ends you can get a reasonable picture. . . . Any new grain will always originate from the outside—from a surface—they won't originate from within the bulk of the material, so if you can effectively draw all the grain boundaries on the outside then you know that—you can see where the actual single crystal area from the boundaries of the grain are.

How the Story Ended

This was how the situation was left while I had to spend some time abroad. The ingot was to be given to another technician to be cut in half and then polished. When, some months later, I next spoke to Draper I discovered that the cutting and polishing had not taken place.

Draper: I sort of got a second opinion on this and they said they thought it ought to go in for a couple of other times. . . . I think we're going to have to do it two or three more times—I guess. The grains are quite large now so I think the chances of getting something big out of it are fairly reasonable. . . . If we've got five or six, there's quite a reasonable chance we might get two or three.

The view was that there were several crystals in the ingot that Draper wanted to reduce to only two or three. Somehow the collective opinion of what was present had changed.

Draper: There's a large . . . crystal running down the center of it. . . . There's probably about eight or so large grains still in there and we're trying to get down to about three. But you don't really improve things the more times you do it. It's that just chance may give you. . . . Once it just happens to give you two or three grains you stop, and you don't bother to carry on at that point.

Months have elapsed since the last attention was paid to the crystal. Partly this was because, as Draper said, "I went away on holiday and when I came back all the apparatus had been dismantled. This is the actual practicality of things." As a result, a classic error was committed. Draper had forgotten which way round the ingot was supposed to be for the next run.

The time taken to do further runs now stretched out so far that I could no longer keep close contact with the history of the crystal. My reports became very much secondhand. It appeared that Draper became somewhat more pessimistic about the state of the ingot but decided to use another technique to improve matters. He decided that there were good crystals at one end and that he would *melt back* to these and then melt forward again so as to use the good crystals as seeds to encourage growth through the rest of the ingot. This is a very delicate operation because if the heater goes too far, or gets too hot, the seed crystals will melt and one will be right back at the beginning. To make matters more complex, because the grain boundary of the best crystal was not square to the ingot, Draper thought better results would be achieved if the molten zone could be persuaded to travel up the ingot at an angle. To achieve this he decided to blanket off part of the heater with insulating cloth. He eventually achieved all this and produced a satisfactory crystal.

Cadmium-Tin-Phosphide

I now became apprenticed to an additional master, a graduate student who wanted to grow a crystal of cadmium-tin-phosphide. Again, the rules of crystal growing as we had understood them up to that point were broken immediately. CRYSTAL was wrong or unhelpful on every occasion where I compared it with laboratory practice.

According to our understanding of crystal growing as it had emerged in the knowledge elicitation sessions, cadmium-tin-phosphide could not be grown in an ampule because the vapor pressure of phosphorous at the melting point of cadmium was too high. Yet my new master had discovered a way of using the ampule method to accomplish his aim. (Others told me that this was only after many, many explosive failures.)

In the closed-ampule method the mixture of chemicals is put into an ampule—perhaps 20 centimeters long and 1 centimeter in diameter and made of glass or quartz—then the ampule is placed in a furnace in an upright or sloping position. The ampule is heated to a temperature well beyond the melting point of the constituents and allowed to cool slowly so that the bottom freezes first and the top last. The way this is done in practice is to heat the whole furnace in such a way that the top is hotter than the bottom and then to cool the whole thing at a uniform rate. That way, the temperature reaches the freezing point of the materials at the bottom first and the zone of freezing moves up the ampule. The process works best with a large temperature gradient between the top and the bottom so that the freezing zone moves very slowly. The temperature gradient is measured by attaching three *thermocouples* to the outside of the ampule, one at the top, one in the middle, and one at the bottom.

The Problem of Vapor Pressure

It was explained to me that with phosphorous the problem of vapor pressure, bad enough anyway, is exaggerated because one requires an excess of phosphorous to make certain that the eventual proportions in the crystal are correct. If the chemically exact proportions are present at the outset, then the final crystal is likely to be deficient in phosphorous. Excess phosphorous is used to force the proper proportion into chemical combination with the other elements. One can see that this makes most of the decimal places on the chemical balance irrelevant, at least so far as the phosphorous is concerned, because one uses an excess of *roughly* 5% to 10%. CRYSTAL's *front end* is capable of providing the required masses of elements to four places of decimals. It would actually be thoroughly misleading to a novice just as my calculator is misleading when it gives me my height in centimeters to four places of decimals. But one would not want less accuracy in the calculations because the judgment of appropriate accuracy has to be left to the user; it varies from case to case.

My new master's method of growing phosphorous-containing crystals in an ampule was to minimize the effects of the phosphorous by condensing a large proportion of it at one end of the ampule, which is kept cold during an initial heating phase. In the meantime, some of the phosphorous will enter into chemical combination with the other elements, eliminating its contribution to the internal pressure of the ampule. The remaining phosphorous can then be heated along with the rest of the ampule in the normal way and most of it will then enter into chemical combination. The remaining small excess of phosphorous will not burst the ampule. This complex method shows that the idea of vapor pressure cannot be translated into pressure inside the ampule. The net pressure inside the ampule is a matter of how the temperature is controlled as well as the relationship between quantity of volatile materials and capacity of the ampule. The new method became inscribed in CRYSTAL in the fall-back method that we called *partial compounding*. Clearly, each new case of partial compounding requires skill plus trial and error to make it work. Solutions cannot be anticipated prior to the event.

In the last chapter I recorded that Draper favored the closed-ampule method because one can

> seal everything up so what's in there is going to be more or less what your crystal will consist of. You know you're not going to have too much problem of things disappearing off and condensing somewhere else.

I remarked that the idea of the ampule as an easily controlled closed universe was misleading. Now we can see why. Even in the small world of an ampule containing three elementary chemicals there are so many variables that the behavior of the system cannot be predicted. An ampule is a small world, but it is far from being the *microworld* of a program like SHRDLU.

Computing and Intelligence

As before, to grow the crystal we had to weigh out the chemicals and load them into the ampule. This time we started, not from Draper's notes on previous runs, but from the chemical formula for cadmium-tin-phosphide, or Cd, Sn, P_2. Given the final mass of the crystal toward which we were aiming we could work out the required masses of the elements from the formula and the atomic weights. We could then add 10% or so of excess phosphorous. This is the sort of simple but irritating calculation in which it is very easy to make a mistake, as my instructor pointed out. I suggested that I would work it out at the same time as him and we would see if we got the same answer. The results of our rival calculations differed only at the fourth decimal place (probably due to calculator rounding errors), and we decided that this level of accuracy was good enough.

I had studied chemistry at grammar school. At last my sixth-form training had found a use. I was delighted that I could do the calculation of masses from my understanding of atomic weights. It would be very easy, however, to delegate that particular calculation to a computer and I decided to program our expert system to do it. I wrote the program (in BASIC) in a few minutes. This little *front end* for CRYSTAL, the bit that does straightforward, regular, old-fashioned computing, is probably the most useful bit of the whole program. It does something that even an expert finds irritating and error prone. Everything else that CRYSTAL does is either wrong or too trivial to be worth bothering with from the point of view of the expert, or wrong and unusable from the point of view of a novice.

A few minutes of old-fashioned number crunching is worth months of PROLOG. This, of course, is because number crunching—being a machine-like activity—is hard for humans whereas the rest of life is very hard for machines. Those who have some experience of programming computers will have noticed that a few lines of BASIC can achieve things that no human can do, whereas the result of months and months of artificial intelligence work can be bettered by almost anyone.

Making a Thermocouple

My next job was to make the thermocouples to record the temperatures at three points of the ampule. The word "thermocouple" brought to mind a technological device, well beyond my abilities to make. Reality was, as usual, demystifying. It also provided a lesson in the practical skills that always lurk just beneath the surface of our technical language.

A thermocouple is simply a pair of dissimilar electrical wires joined at the end. They generate an electrical signal if they are heated at the join. To make a thermocouple one starts by covering two lengths of wire with hollow ceramic insulating tubes. Each tube is about 4 centimeters long and 3 millimeters in diameter with a small hole through the middle. One has to thread these elongated beads onto the wire until about a meter of each kind of wire is encased. Then the ends of the wires are twisted together and fused in a blowtorch flame. This makes one thermocouple. The hardest part is threading the ceramic tubes onto the wires. Draper told me how to do it:

Draper: Try and keep it [the wire] straight, otherwise you'll find it difficult to feed up the middle of these little insulator tubes.

Collins: Bob, as a matter of interest, do you usually put one up at a time?

Draper: I usually put them up in pairs.

Then I fused the ends of the wires under my new instructor's guidance:

Collins: Like that?

Instructor: Yes, that's fine. Just do it very, very slowly. Put your arm here—same hand or you can't control it. Move it down more—more, more, more.

Collins: Down here?

Instructor: Yeah. Go very, very slowly to the fire. Very slowly. A little bit more—go, go, go. That's fine. That's alright.

And a little later I learned that a furnace was a homemade box filled with insulation and heating wires and that a *two-zone furnace* was the same thing with a hole in the middle. And I learned that when one first melts up an ingot in a furnace it is important to make sure that there are no air gaps in the mixture and that this is done by switching on a mechanical vibrator attached to the frame of the furnace but, more important, by bashing the frame very hard, very loud, and very long with a pair of old pliers that happen to be lying there. "This," I remarked to myself, "is real physics."

And thus, and thus, did I begin to learn to grow semiconductor crystals. I still have a very long way to go.

Conclusion to Part III

The last three chapters have been intended to show the difference between abstracting rules of practice for use in a text or an expert system and learning the practice of the same craft. They have shown just what it means for rules not to contain the rules of their own application. They have shown what it is that a user brings to the interaction with an expert system. Readers might like to think about just how little they could achieve using CRYSTAL unless they already knew their way around a laboratory and they already knew how to grow crystals. In fact, we tested CRYSTAL out on some fairly skilled users. They all loved to play with it—to an extent that surprised us.[4] They liked to find their way around the rule base by using the "how's" and "why's." They felt that there was a class of user (not themselves) who could learn something of crystal growing from such an exercise. One visitor even took a copy back to Turkey. Perhaps CRYSTAL has a role in teaching people the elementary rules of crystal growing in an amusing way.[5] But, front end aside, it is hard to see how it could ever be put into day-to-day use by an expert or a novice.[6]

I have looked at three descriptions of semiconductor crystal growing. The first of these was the model extracted from our expert and encoded in CRYSTAL. We first thought of this as a model containing practical experience—the sort of practical experience that is not found in textbooks. We contrasted the practical techniques in CRYSTAL, such as the ampule method as a solution to the problem of pressure, with the second description—the textbook. This was comparatively impractical. I showed how respect for textbook knowledge made it difficult to extract technical tips from our expert.

The third description was that of the apprentice. In the laboratory even the practical tips encoded in CRYSTAL turned out to be a misleading mask for a set of special cases. Our painful knowledge elicitation sessions, though they brought us a long way from the textbook, had not brought us very near to the workfront.

Interestingly, two opposite effects began to unfold as I worked in the laboratory. On the one hand, if I try to describe my experience in detail, the description becomes more and more ramified; on the other hand, the process of crystal growing becomes demystified and simplified as it is encountered. The actual common sense as it becomes *my* common sense, gets simpler and simpler—less and less forbidding and mysterious—and, of course, less and less visible. It becomes less and less a thing that I consciously possess and more and more an indescribably familiar part of me. Both the indefinite ramification of description and the folding-in of experience are fatal for the project of knowledge engineering understood as the task of replacing experts with machines. They are both benign so long as expert systems are machines for helping experts.

IV

Testing the Limits of Articulation

13

The Meaning of the Turing Test

The Test in Outline

In 1950, *Mind* published a paper by A. M. Turing entitled "Computing Machinery and Intelligence." In it, Turing proposed a way of answering the question "Can machines think?" What has become known as "The Turing Test" is an operational answer to this question. It rests upon the machine's ability to imitate a human in typewritten interchanges. Turing's initial formulation of the test was based on *the imitation game.* In the imitation game a man and a woman are concealed while an interrogator tries to determine which is which by asking written or typed questions. To play the game the concealed participants both agree to reply as, say, women, with the concealed man pretending to be a woman while the woman does her best to convince the interrogator that it is really she. Turing suggested that the man be replaced with a machine.

> We now ask the question, "What will happen when a machine takes the part of [the man] in this game?" Will the interrogator decide wrongly just as often when the game is played like this as he does when the game is played between a man and a woman? These questions replace our original, "Can machines think?" (Turing 1982, p. 54)

A Subtle Feature of the Turing Test

The Turing Test looks simple but has hidden subtleties. The first thing to notice is that the machine is to mimic a man who is imitating a woman—not a man being himself. It is far easier for a machine to mimic a man pretending to be something he is not than for a machine to mimic a man acting authentically. Like the spy in Semipalatinsk, a man pretending to be something he is not will have a far smaller repertoire than normal. What is more, the repertoire will be of the sort that one can learn through formal teaching rather than socialization. This is an ideal repertoire for teaching to a machine. Turing's test might have been designed deliberately to conceal the contribution of socialization to the formation of human abilities.

Interestingly, it is a social change that has altered the test so that this subtle feature of the design has faded away. In Turing's day the sexual imitation game was different. In the 1950s one would have expected a clever interrogator to be able to detect the concealed man on substantially more than 50% of occasions. Men, in those days, did not share enough of the life of women to mimic them easily.[1] That is what made it easy for the machine to replace the man in the test. Nowadays men and women have much more in common. In some informal experiments conducted with undergraduate students at the University of Bath in the mid-1980s, we found that the frequency with which the students could tell the sex of respondents in the imitation game was almost chance. Because of men's and women's common experience the game is no longer exciting. Men and women are no longer culturally separate groups in the way that Londoners and Semipalatinskians are. Nowadays a machine imitating a man imitating a woman has a task as hard as a machine imitating a man being himself.

It is easy to regain the sense of Turing's original design by replacing men and women with groups that are culturally diverse today. Ask, "What will happen when a machine takes the place of an English spy in Semipalatinsk? Will the machine do as well as the spy?" We can understand that to mimic a spy is to mimic only a narrow range of human abilities; culture has been written out of the definition of intelligence.[2] To adapt the Turing Test to our purpose we need culture to play its part. The first step is to eliminate the requirement that the replaced human be mimicking someone else.

The Protocol

In its simplified form the test requires three separate spaces or their conceptual equivalent (see figure 13.1). A human control, *h*, goes into cell 1 (or 2) and the machine *m*, which attempts to imitate the responses of a human, is placed in cell 2 (or 1). The interrogator, who cannot see into the cells, communicates with *h* and *m* through two identical channels (usually linked teletypes), *a* and *b*. The interrogator is to distinguish between *h* and *m* by the differences in their conversational ability. Insofar as the machine is indistinguish-

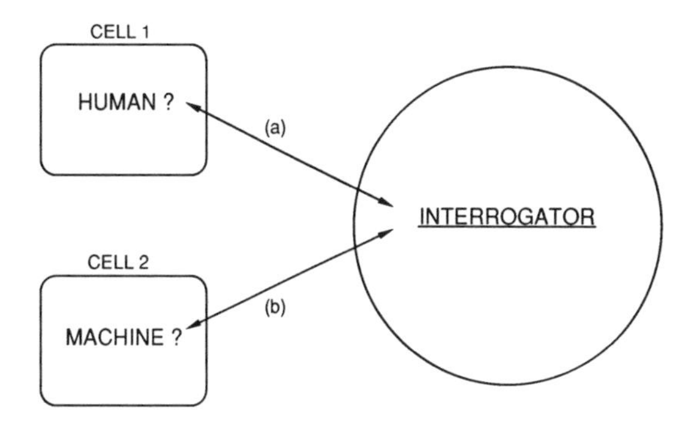

Figure 13.1
The Turing Test

able from the human in this test, then it can be said to have passed the test and to be a thinking machine.

The Turing Test is so interesting because it is a test of the capacity of a machine to mimic the *interactions* of a human being. It is, then, a test of our abilities to make an artificial human that will fit into a little social organism as opposed to a test of a machine's ability to mimic a brain. What is more, the appearance of the machine is of no consequence, and this again fits the analogy of an artificial heart rather than an artificial brain (see chapter 1). Of course, the machine operates in restricted circumstances that give it every chance of success. This has been taken to be a defect by some writers:

> Computers lack that "rest of behavior" . . . [Wittgenstein 1953, para. 344] which is one of our criteria for the ascription of "inner" phenomena like "intelligence"; in this sense, they do not take part in the bustle of human life. Therefore, we cannot decide upon the possibility of artificial "intelligence" by tests like Turing's that abstract from this context. (Neumaier 1987)[3]

I believe, however, that making the task easier for the machine and harder for the critic makes for a much more interesting philosophical challenge. In any case the Turing Test protocol is not too dissimilar from the social interactions we have with, say, telephone switchboard operators. If a machine could mimic a switchboard operator then one would want to call it intelligent.[4] Thus, in spite of Neumaier's reservations the engineering question about artificial intelligence is nicely approached from the perspective of the Turing Test. It is a *hard case* for the skeptic's argument—the only kind of case from which generalizable conclusions can be drawn.

First Sociological Thoughts about the Turing Test

There are other subtleties in the way Turing wrote about the test. He was far more careful in his claims than most of those who use the test in their arguments. He said:

> I believe that in about 50 years time it will be possible to programme computers . . . to make them play the imitation game so well that an average interrogator will not have more than 70 percent chance of making the right identification after five minutes of questioning. The original question "Can machines think?" I believe to be too meaningless to deserve discussion. Nevertheless I believe that at the end of the century the use of words and general educated opinion will have altered so much that one will be able to speak of machines thinking without expecting to be contradicted. (Turing 1982, p. 57)

With thirty-eight of the fifty years elapsed the last sentence seems like an accurate, even conservative, estimate of progress. As for the first sentence, a number of claims have already been made to the effect that programs have "passed the Turing Test." But it all depends on what you mean by "the Turing Test" and what you mean by "pass."

Turing talks of an average interrogator. Linking this with the final sentence makes it clear that he thought of the test not as a carefully designed and controlled laboratory experiment but as a broad statement about likely trends in the development of computers and their impact upon the general public. Thought of in this way, details of the test's experimental protocol are not salient, and the extent to which Turing's predictions will be fulfilled or surpassed has only a little to do with the development of computers and much more to do with the way we think about them. In this Turing could be said to have anticipated the relativist approach of modern science studies. Claims that programs such as ELIZA, or PARRY (Colby et al. 1972), have already passed the test are better seen as claims about the proper way of talking and thinking about computers than claims about the success of carefully designed experiments.

The Experimenter's Regress

This way of looking at the matter grows naturally out of recent work in the sociology of scientific knowledge. Consider the *Experimenter's Regress* (Collins 1985). The idea of the Experimenter's Regress starts with the observation that experiments are difficult and need a great deal of skill. What is more, it is impossible to know whether one possesses the ability to do an experiment without actually trying it out (as with other skills such as riding a bicycle). This means one only knows if one can do an experiment by trying it, observing the result, and seeing if this result is in the correct range. Sometimes the experimenter does not know in advance what the correct range is and then it is not possible to be certain that the experiment has been done properly. In the natural sciences problems arise in experiments that try to detect disputed phenomena. To know whether "thargs" exist one must build a good tharg detector and look. But one can only determine if one has a good tharg detector by exposing it to thargs and observing the outcome. But one can only know if thargs are there to be found by building a good tharg detector and looking for them. But to know whether one has a good tharg detector one must know whether thargs exist and so on ad infinitum. Discovering (or discovering the nonexistence of) thargs is a process exactly coextensive with finding out how to build a good tharg detector.

If the Turing Test were ever to become a serious laboratory experiment the type of arguments that would follow are predictable. People would argue in the same way as they do over, say, card-guessing experiments in parapsychology. In card guessing a "subject" takes a set of shuffled cards and looks at each in turn while a psychic, who is sensorily shielded from the subject, names the cards as they are viewed. If the psychic makes correct identifications at a level significantly higher than chance, then the hypothesis of *extrasensory* perception is taken to be confirmed. Large numbers of these experiments have been done and many have significant positive results. Critics, however, refuse to accept these results at face value maintaining that the experiments must have been flawed. Critics think of ingenious ways in which the psychic may have circumvented the protocol to gain *sensory* perception of the cards. Given ingenuity on both sides, the argument about an adequate protocol for such a test is potentially endless. Each design improvement made by the parapsychologists is matched by the critics' invention of a possible new ordinary-sense channel of communication.[5]

In the same way we can imagine proponents of artificial intelligence, finding that interrogators are easily able to tell the machine from the human in Turing Tests, searching for channels of communication between interrogator and machine other than the teletype link. Is the machine humming in such a way that it can be detected subliminally? Is the human control sending a prearranged code down the teletype line? Are there collaborators spying on the test building and transmitting the true location of the machine via a radio embedded in the tooth of the interrogator? The contrary case—where the interrogator finds it impossible to tell machine from control—is even easier to deal with. The interrogator might be incompetent; or the interrogator might be biased and fake his or her inability to distinguish machine from human, or the machine might be rigged by a cunning technician to bypass the program and transmit messages from another concealed human. There is no need to invoke deliberate fraud in order to question the protocol; incompetence, accidental cues, or poor design for detecting the *essentials* of intelligent behavior are all arguments that are likely to be used as they are used throughout contentious science.

Discovering the *correct* protocol for the Turing Test is a matter of making certain that the test identifies intelligence in a machine only if it is really there; but we can know if our test works properly only by trying it out on a genuinely intelligent machine; but we can know if a machine is intelligent only by running a Turing Test on it. The definition of intelligence is intimately related, then, to the definition of the correct protocol for determining its existence.[6]

Alternative Test Protocols and the Definition of Intelligence

Given this way of thinking, Turing might have designed the test to make it easy for computers to be defined as intelligent while intelligence became defined as what computers exhibit in Turing Tests. Searle's or Neumaier's type of argument try to make intelligence harder to accomplish and, inter alia, to define it as something entirely beyond Turing's protocol. So long as the Turing Test continues to be counted as an indicator of intelligence, however, then so long will there be explicit or implicit attempts to redefine its protocol so as to suit the preferred definition.[7]

Consider some other explicit and implicit definitions of proper test protocols. When it is claimed that ELIZA passes the test because some people are as happy with it as they would be with a human psychiatrist, the implication is that the human control need not be present at the time of the test. People who think ELIZA is as good as a human are comparing it with their *idea* of a psychoanalyst rather than with a particular psychoanalyst who is there at the time of the consultation. Further, the implication is that the interactions typical of nondirective psychotherapy are an appropriate *subject matter* for the test.

The same applies when Colby's (1981) experiment with his paranoid schizophrenic program PARRY is said to reveal the capacity of machines. In this experiment judges were asked to diagnose patients from an interview transcript. It turned out that they were indifferent between human patients and a computer program intended to simulate the responses of a paranoid patient. Furthermore, psychiatrically trained judges, given the transcripts of real paranoids and the transcripts produced by the program, were able to do no better than chance when asked to identify the source:

> 100 randomly selected psychiatrists were sent protocols of two interviews, one with a patient and one with a program, and were asked to judge which was which. Out of 40 responses 21 (52%) were right and 19 were wrong. Since this approaches a chance level, we can conclude that psychiatrists using teletyped data do not distinguish real patients from our simulation of a paranoid patient. (Colby et al. 1972, p. 220)[8]

But is a paranoid personality a suitable subject for a Turing Test or is it too easy to mimic? Joseph Weizenbaum (1974) said that paranoia, like other mental pathologies, was too easy. He remarked sarcastically that autism could be mimicked by a typewriter just sitting and humming. Is mimicking a paranoid schizophrenic a demonstration of a program that can pass the Turing Test, or is it better seen as an implicit argument about the test's protocol—an argument that makes intelligence an easy thing for a machine to accomplish? What is the meaning of the following quotation: "Of course, today's computers can already fool experts and most computer scientists now think the Turing test is too easy." (Shurkin 1983, p. 72)

Another vital question is whether the interrogator, or judge, engages in the interaction in the spirit of a test or in a more respectful frame of mind. Those who have cited ELIZA's interactions as passing the test have implicitly redefined the protocol so that the interrogator does not have to be aware that a test is going on. This is especially important because we are so able and willing to repair human interaction if we do not suspect we are the subject of a trick. The point is made dramatically and convincingly by Garfinkel's counselor experiment described in chapter 7. Garfinkel, it will be recalled, found that students were able to make sense out of yes or no answers to queries about their personal life even though the answers were given at random. The students, we may be sure, would not have been satisfied with the answers if they had *known* that a table of random numbers was responsible. The crucial question is: "Would they have been able to distinguish the table of random numbers from the answers of a real counselor if they had been asked to distinguish between the two?"

Colby did find that judges conducted their interviews differently if they thought a machine was involved:.

> What a judge is told, and what he believes is happening, are extremely important in indistinguishability tests. From our experience we have learned that it is unwise to inform a judge at the start that a computer simulation is involved because he then tends to change his usual interviewing strategy and ask questions designed to detect which of two respondents is a program rather than asking questions relevant to the dimension to be identified. (Colby et al. 1972, p. 202)[9]

These different features of Turing Test protocols must make a difference to what we think of as intelligence and what we think of as *the test.*

Computers as Part of Our Culture

Now go back to the last sentence of Turing's quotation—by "the end of the century the use of words and general educated opinion will have altered so much that one will be able to speak of machines thinking without expecting to be contradicted." One might reasonably take being "able to speak of machines thinking without expecting to be contradicted" as a good indicator that the thinking machine had entered our "form-of-life"—an indicator that is distinct from questions about what the machines can actually do. How else is this kind of talk and thought encouraged? Consider the book title *Machines Who Think* (McCorduck 1979). The more we use computer metaphors to describe ourselves and the more we allow the computer to define the limits of our thoughts and studies, the more readily will we think of computers as like ourselves.[10] As this happens it will become more and more easy for a machine to pass a Turing Test because computer-like responses would become what we expect of humans. As Dreyfus (1979, p. 280) puts it in a colorful passage:

Man's nature is indeed so malleable that it may be on the point of changing again. If the computer paradigm becomes so strong that people begin to think of themselves as digital devices on the model of work in artificial intelligence, then, since for the reasons we have been rehearsing, machines cannot be like human beings, human beings may become progressively like machines. During the past two thousand years the importance of objectivity; the belief that actions are governed by fixed values; the notion that skills can be formalized; and in general that one can have a theory of practical activity, have gradually exerted their influence in psychology and in social science. People have begun to think of themselves as objects able to fit into the inflexible calculations of disembodied machines: machines for which the human form-of-life must be analyzed into meaningless facts, rather than a field of concern organized by sensory-motor skills. Our risk is not the advent of superintelligent computers, but of subintelligent human beings. [11]

Novelty, Mystery, Plausibility

Another way of thinking sociologically about the relationship between ourselves and computers is to seek precedents in our relationship with other technological artifacts. For example, Searle (1984) has pointed out that we tend to attribute intelligence to whatever has been newly invented. Thus at one time telephone exchanges were taken as a model for the human brain but as they became more familiar and new technical devices such as computers were invented the model of intelligence moved on. Close to this is the idea that we attribute special powers to things that are mysterious but cease to attribute them once the mechanism has been understood. Woolgar (1985) illustrates this nicely by asking us to think of an intelligent television that can switch off the advertisements. He suggests that we would think of such a device as intelligent only so long as it seemed to understand the content of message. We would cease to think of it this way if we learned that it worked by, say, responding to a change in signal level that always corresponded to the change from program to advertisement.

One could say part of the force of Searle's Chinese room argument rests on this kind of idea, albeit in an implicit way. He says that even though the person with the code book would pass the test we do not count the Chinese room as understanding. By definition the room performs as well as a Chinese speaker, so our unwillingness to attribute understanding can only be because we know what is inside the cell. Though, on the face of it, Searle's argument is analytic—he points to the sense in which the non-Chinese speaker does not understand in the same way as a Chinese speaker—it really rests on our disinclination to attribute intelligence to certain kinds of mechanism. Searle shows that familiarity with the mechanism of the Chinese room leads us to be less inclined to locate it alongside us in our forms-of-life.

This way of looking at things fits well with a far more general idea that has already been touched upon: close examination of a passage of activity is demystifying (while distance from the seat of creativity lends potency to the created product).[12] The trouble is that disenchantment is not easy to bring about because it is hard to bring people close to sites of creation. Indeed, the opposite, anthropomorphic, force is the stronger one. In the main we read the *meanings* of artifacts as their *use.* If computers, or Chinese rooms, really worked as well as Searle hypothesizes then we would start to use them in the same way as we use humans. They would become located alongside us in our networks of practice. And because practice and knowledge are but two sides of the same coin, other things being equal, we would probably start to think of them as us and us as them.

A Turing-type test is a good test for computing ability precisely because it is a test of the extent to which a machine can be located in an interactional network without strain. Instead of asking about the innate ability of the machine one looks at its interactive competence; this is how we judge other things that interact with us. For example, we do not leave judgments of other humans in abeyance until we have examined them for hidden mechanisms nor do we necessarily care about such things as pacemakers, artificial hearts, or bits of metal embedded in the brain so long as the people act in a normal way. The androids of science fiction are a cause of concern only when they go out of control. That is, they worry us only when they cease to fit neatly into our sociocognitive networks. On the other hand, insane humans are a cause of concern because they do not fit neatly into our society. How they work is not the problem. At the same time, we are ready to anthropomorphize our animals and even our machines when they are sympathetic.[13] This suggests that if machines come to have the same use as us in large parts of our practical life, it will need work and vigilance, education and propaganda, to keep them separate from us (assuming that is what we want to do). It is that kind of intervention that I now make in defining the proper protocol for the Turing Test. This means moving away from relativistic sociological analysis in order to participate in the debate.

The Turing Test and Intelligence

Where a prosthesis such as an artificial heart or a social prosthesis such as Garfinkel's counselor is deficient, the body or social group will compensate. In the case of intelligent machines the user must cooperate. We can now ask the *ultimate engineering question*: "Could a machine pass a Turing Test not only in the absence of cooperation but while being interrogated by someone trained to detect machines?" To answer the question we have to think very hard about what the ideal protocol for the test would be and how the interrogator would be trained. The plot is drawn from science fiction and from the world of spies. Can a machine pass as something that it is not under hostile interrogation? How can we learn to trap these strangers in our midst?

Because we are interested in socially acquired abilities we will restrict the protocol to make these the focus of attention. We are not interested in what a detective could discover by physically dissecting the social prosthesis—that is a question for a different sort of engineer. Nor, at this stage of the art, is it interesting to refer to the machine's appearance, mobility, or physical skills. Given these constraints, thinking about how we might train our interrogator and how we might design the test *is* thinking about the limits of artificial intelligence and the nature of intelligence itself. Reaching a conclusion is staking a claim about what intelligence is.

Protocol for the Ultimate Turing Test

Turing talked of the *probability* of a machine passing the test. He looked forward to a time when interrogators would have only a 70% chance of telling the difference between machines and humans in a five-minute test. This might be an uninterestingly easy task for the following reason. There are some human beings whose responses in a Turing Test-like setting would be more like a machine than even ELIZA's. Weizenbaum's remark about the autistic person and the typewriter makes the point, but it also applies in more subtle circumstances. Imagine something like the Chinese room being subject to a Turing Test but let the interrogator and the control have learned Chinese only from books. Under these circumstances a fairly elementary Chinese room could pass the test. There would be no cultural dissimilarity between any of the parties. In this case neither control nor interrogator would understand Chinese in the sense in which Searle uses the term. Even a human executing a task does not necessarily have to understand in order to execute the task. It can be performed by rote.[14]

If there are a high proportion of humans who are more machine-like than human-like, a fairly middling program will be likely to be misidentified when matched against a sample of the general population. It will, then, be very easy to get down to the 70% probability mark, or even to 50% or *below.* [15] To save embarrassment we must say that the ultimate engineering question is about mimicking the full range of human performance rather than an easy subset. If the full range of human abilities is to be mimicked, then the most human-like person must be mimicked just as satisfactorily as the least human-like. The interrogator must have no better than a 50% chance of correctly identifying the most human-like person there is. Likewise, the interrogator will need to be the best there is at detecting human-likeness. Our question then becomes, Is there is a control-interrogator pairing, the human-likeness of whose interactions cannot be mimicked by a control-machine pairing, given an ideal protocol? What can we say about a perfect control-interrogator pairing under a perfect protocol? What does perfection mean in this context?

Let us work through each critical feature of the test in turn, reiterating points already covered as well as adding new ones.

1. The attitude of the interrogator: Does the interrogator know that it is a test situation? Will the interrogator be charitable or skeptical?[16]

Colby, the psychiatrist who invented the paranoid program, is probably unique in actually conducting a carefully designed Turing Test-like experiment rather than simply discussing it in theory. Nevertheless, his experiment suffered from the deficiency that interrogators did not know that one of the conversational partners was a machine (Heiser et al. 1979). His experiment did have a second stage in which judges who knew the nature of the test failed to distinguish between transcripts of interviews with paranoids and interviews with PARRY, but the transcripts had been produced by interviewers who were ignorant of their intended purpose. Therefore none of the questions asked by the interviewers was meant to separate machine from human. Such a protocol has no bearing on the ultimate engineering question. To be relevant the test must be conducted uncharitably, by an aware interrogator, in real time, so that there can be creative interaction.

2. The subject matter to be discussed in the test

A test limited to specific subjects might be very easy to pass. A machine equipped with a "script" pertaining to a narrow domain might do well on such a test. A test limited to arithmetical questions with all interaction expressed in arithmetical notation would require an especially vigilant interrogator. Imagine a human arithmetician competing with a calculator in the mathematical room of chapter 4. If the interrogator had been trained by reading chapters 4 and 5 of this book, then he or she would have a good chance, but otherwise the subject matter would make the test too easy for the computer. The general principle is that the subject matter must not lie within the domain of machine-like action.

3. The length of the test and the number of iterations

If the test lasted only 20 seconds there would almost certainly be insufficient time to get a response. A block of wood could pass the test under these circumstances. Tests that last long enough for only a few short interchanges could, other things being equal, be passed by very simple programs. ELIZA usually puts up a good show for a few interchanges unless the respondent is extremely sophisticated.

Another reason to be concerned about the length of the test is that short tests might be passed by machines that rely predominantly on memory whereas longer sessions would reveal repetitiveness and narrow scope. The flaws in certain programs would be revealed only by stopping the test and starting again from the beginning. Some programs would repeat identical responses to identical questions.

There is an analogy with random numbers. If one has never seen a particular table of random numbers each new number is unpredictable but second time around the numbers, though just as random according to internal checks, are completely predictable. Imagine a Geiger counter confronted with a small quantity of radioactive material. The intervals between its clicks are random. Now imagine a "clicking machine" driven by a program based on a table of random numbers. A judge would not be able to tell which was the natural process and which the artificial. Second time around, however, and the random number sequence would repeat itself whereas the Geiger counter would still be fundamentally unpredictable. Likewise, in human interaction one would expect some variation in responses between runs. With this in mind, programs (like ELIZA) that repeat responses to questions after one or more runs might be said to mimic only "quasi-human" interaction.

To be an adequate test, then, the protocol must allow not only a long test period, but also the possibility of repeating the test a number of times from the beginning so that quasi-human responses can be detected.

4. The ability, knowledge, and behavior of the control *h*

The ideal control in the most stringent Turing Test protocol would be someone who was maximally human-like. It is hard to say what the essence of human-likeness is although that is what we are groping toward. What we know is that it will have something to do with what can be learned from socialization and something to do with making what will be seen as proper responses under unanticipated circumstances. In the next chapter I argue that these can be recognized in the ability we have to repair others' interactional mistakes. Clearly human-likeness has nothing to do with machine-like acts such as the central parts of arithmetic.

We can eliminate some totally unsuitable types of control at the outset. An unsuitable control would be someone who was unable to manage conversational responses, or whose range of responses was unusually limited. That is why a Rogerian psychotherapist would not be a suitable control. It partly explains the success of ELIZA and PARRY. Even if the control has "normal" human competences, certain behavior may make it a trivial task for the computer to pass the test. Colby et al. (1972, p. 203) found that "if a human-respondent does not follow standards of the interviewer's expectations, jokes around, or plays other games, ordinary judges cannot distinguish him from a computer program with limited natural language understanding." Just how true this is must depend on the nature of the games and jokes. A certain kind of responsiveness to jokes heard for the first time is exactly what we might not expect from a computer, whereas unsolicited joking around would be easy for a machine to mimic. The best possible control would be someone who can exhibit the full range of human responses to the interrogator.

5. The abilities and training of the interrogator

Someone with a restricted conversational repertoire would not make a good interrogator for the same reasons as they would not make a good control. A block of wood could pass the test if the interrogator did not know how to ask or understand the answer to a question. More significantly, the interrogator must know how to exploit the conditions of the test. The question is whether machines can mimic the full range of human abilities when confronted by the cleverest and most well-trained inquisitor.

There is no straightforward recipe for training an interrogator. There is, I believe, one overarching principle, which turns on the ability of humans to repair deficiencies in others' talk; I introduce the principle at the end of this chapter and develop it in the next. This apart, every type of interaction needs to be thought about separately. For example, very good chess players could almost certainly work out whether they were playing against a human or a machine provided they had some idea of how the machine worked so that they could look for typical machine moves. I could not tell which was which. Chess, of course, is a case that is heavily loaded in favor of a machine because the moves are digitized. (Played via a teletype, few of the flexible elements of the meaning of chess, discussed in chapter 2, are relevant, though perhaps the current generation of machines would not know when it was culturally appropriate to resign.) We have talked of the mathematical room and have touched on another case—automated spelling checkers. As a result of the earlier discussion of spelling checkers, readers should know enough to tell a human spelling checker from a machine by looking for relative differences in the ability to cope with correctly spelled mistakes of sense (such as "weird processor"), and small mistakes in long, complicated words.

6 The relationship between control, machine, and interrogator

Think back to the male-female imitation game upon which Turing modeled the test. Distinguishing between men and women is hard these days because men know so much about women's lives. But, given this, a woman interrogator would have more chance of detecting the residual mistakes made by a fake woman than would a man and vice versa. Men are less able to distinguish between a real woman and a man pretending to be a woman because the responses a man expects of a woman will be what he would give if *he* were pretending to be a woman. That is, the man interrogator's model of a woman will be similar to the fake man's model of a woman (deficient in similar ways), whereas a woman interrogator's model of a woman will be more accurate than that of a fake man. Thus, under these test conditions, the hidden man's female impersonation would seem natural to a male interrogator but not to a woman interrogator.[17] What follows is that the closer the cultural relationship between interrogator and control/machine, the harder it is for the machine to pass the test.

The significance of cultural similarity has been encountered before. For example, I argued that a spy who tried to pass as a native of Semipalatinsk could be unmasked by a native but not by a stranger to the town. The necessity of a close cultural relationship between interrogator and test subjects also shows that the Turing Test could not be used to decide if Martians, dogs, or lions were intelligent; none of them could be expected to interact normally. As Wittgenstein remarked, if a lion were to speak, we would not understand what it said.[18]

To sum up, for a maximally demanding protocol, the control, the target which the machine is to mimic, and the interrogator must be culturally indistinguishable.

7. Channels of communication and *interpretetative asymmetry*

Because there exist no machines that can understand continuous natural speech with any fluency, the channels of communication will have to involve the typewriter keyboard. A machine will have difficulty coping with poorly typed input. Colby solved the problem of spelling errors by having input monitored and corrected before passing the messages on; however, "When the patient interviewed was an actual human patient, the dialogue took place without a monitor in the loop since we did not feel the asymmetry to be significant." (Colby et al. 1972, p. 208). This comment is very important; it relates to what I will call "interpretative asymmetry."

Interpretative asymmetry is the one-way process in which we repair the defects of machines' interactions while machines cannot repair the defects in ours. The ability to repair is learned as one becomes socialized—it is a skill that cannot be fully explicated because it rests on an understanding of what is culturally acceptable. It is not the sort of technical problem that spelling checkers can deal with. Sometimes humans make mistakes in their repairs—something that the game of "Chinese whispers" exploits—but in general these are not the sort of mistakes that computers make. The implication is that there is potentially a very large class of control-interrogator pairs that would make the Turing Test hard to pass. This class includes all those interrogators who have been trained to input conversational turns that require repair and all those controls who can effect culturally reasonable repairs to faulty input such as misspellings. A reasonably aware interrogator looking for this deficiency ought to be able to tell a machine even from a dull "wooden" human. I deal with this point further in the next chapter.

Conclusion

It has been claimed that certain programs have passed the Turing Test. These claims are best read as bids to define intelligence in certain ways. The protocol of the test needs very careful attention if we are to answer the ultimate engineering question: "Is there a human being that cannot be mimicked by a machine under hostile interrogation?" I have discussed a protocol that, in spite of the reliance on typewritten interchange, keeps cultural abilities in the forefront of the definition of intelligence.

14

Skill and the Turing Test

In the first chapter I discussed the predicament of a spy who had learned everything he knew about a strange place from explicit instructions but was to be cross-examined by a native. I argued that because there is an indefinitely large number of things that a native can say about his place of upbringing the spy could not have been told everything. I also argued that the spy would not be in a position to infer everything that he had not been told from what he had been told. In this chapter I will explore the imaginary outer limits of what we could say and do by virtue of being told things and compare this with what we can say and do by virtue of our socialization.

My argument is based on a thought experiment. In this thought experiment I relax the constraints imposed by the typical lifetime of a human being and even the size of the universe. I allow the spy, or equivalent, to be told everything that it is possible to tell in English in a finite time. Though this bizarre thought experiment has no direct bearing on any foreseeable computer program it is, I believe, revealing of the nature of skill and socialization and the predicament of the spy.

The thought experiment starts with an idea discussed in Block (1981). Block describes an imaginary machine that, he claims, can imitate the conversational responses of any human being in a Turing Test of finite length. The design of the machine is based on the fact that the number of typed strings of symbols that can be exchanged in a test of fixed length is finite and the number of possible typed conversations is a subset of these strings. Therefore, in principle, all possible conversations can be memorized in advance. I generalize this idea in order to find out if there are, in principle, any limits to memory-based imitation of human socialization. I allow the machine to have a memory as large as desired so long as it is finite. The machine has to perform as well as a human in the *hard case* (for the human) of the Turing Test. One might say one is testing the ultimate plausibility of Searle's Chinese room hypothesis.[1] My version of the machine is a little more ambitious than Block's model and is equipped to pass Turing Tests with more stringent protocols than Block had in mind.

To describe the machine in our language, it is designed to reproduce the behavioral coordinates of conversational action. It is like a record-and-playback device for conversational interchange. Though every conversation that is put into the machine's memory must be made explicit, knowledge of sentence structure, conversational turn-taking, and so forth—the things we acquire without conscious effort—are entered without needing to be explicated. If human conversational competence could, in principle, be completely reproduced by storing only its behavioral coordinates, then it seems to follow that there is nothing in a skill that cannot be reproduced by finite, if very large, ramification of a rule base. This point of principle is worth exploring even though it tells us nothing about the actual human organism and little about the practicalities of writing programs. At first sight the programmer I describe *does* seem to mimic all our tacit conversational skills. Its residual failures are, however, the illuminating thing.

The Block Machine[2]

In a Turing Test the parties interact via typewriter keyboards. Imagine that the test is, say, one hour in length. Ignore the control for the time being and let all the interaction be between the interrogator and the machine. If they both type fast, they might type 20,000 characters between them in an hour. Call such a collection of 20,000 symbols a "symbol-string." If the dialogue is in English, each character will be one of about 100 possibilities—26 letters of the alphabet, their capitalized equivalents, numbers, various punctuation marks and other miscellaneous symbols, the space, and a return symbol to signify the end of a conversational turn. Thus, given an hour, the number of different strings of characters that might have comprised the interchange—that is, the number of symbol-strings that there are—is about 100 raised to the power of 20,000, or $10^{40,000}$. This is a finite number, albeit a *Very Very Large Number* (the number of particles in the known universe is about 10^{125}. One can adapt the argument for a test of any finite length.

Most of these strings would not be sensible conversations. For example, there would be one comprising 20,000 *a*'s, one comprising 20,000 spaces, one comprising a long passage from "A Midsummer Night's Dream" but with the word "frobscottle" in the middle of one of Titania's speeches while another contained the word "fribscottle" instead. Another excruciating string would make complete sense except that the very last letter was an *f* instead of an *e*, and so forth.

Some of the symbol-strings would consist of sensible conversations. Chance would guarantee that in some strings all the spaces, letters, punctuation marks, and return symbols would be arranged to make a sensible English interchange. Call such a string a "conversation string." Each conversation string consists of a series of *turns* separated by the return symbol. Every possible conversation string would be included somewhere in the Very Very Large Number of strings in the complete set of combinations of symbols. That is, everything sensible that it is possible for two people to type to each other in an hour would be included somewhere (mixed up with the nonsense strings). For example, all possible hour-long quotations from *Alice in Wonderland* and Shakespeare, all the songs ever sung by Liverpool Football Club supporters, everything that Edward VIII said to Mrs. Simpson, and Profumo said to Christine Keeler and Mandy Rice-Davis would be included, and so would everything you said between 11 P.M. and midnight last night. There also would be conversation strings representing all hour-long conversations that have never happened but that have the potential to take place so long as the basic set of symbols in the English language remains the same. There is nothing you or I can type that is not included somewhere in the set of conversation strings.

Now, suppose we generate every possible symbol-string by some mechanical means and type them all out. We hand the printed list to a *programmer.* The programmer is told to discard all the nonsense strings, extracting just the conversation strings that make sense. He or she (hereafter, "he") has to winnow out the conversation strings from the symbol-strings.

Now, selecting sensible strings is a complicated business. The programmer will not extract every conversation string, because he will not recognize certain strings for the sense that they are. The strings that the programmer extracts will depend upon what *he* counts as a reasonable interchange. For example, if he is not a philosopher, he might miss some of Wittgenstein's "Remarks" for they would not seem to make sense. Likewise, some programmers—those not familiar with A. A. Milne—would miss the meaningfulness of the term "Pooh Sticks" whereas others would not recognize that passages of, say, *Studies in Ethnomethodology* comprise sensible interchanges in the English language. Even if the programmer ignores all strings that contain turns that he does not recognize as sense, the number of conversation strings that a typical programmer would select from the Very Very Large Number of symbol-strings would still be much larger than the number of particles in the universe; the number of strings in the average programmer's set of selected conversation strings will be a "Very Large Number."

The Very Large Number of conversation strings is then encoded into the memory of the Block machine. (It should be clear by now that this can never be more than a thought experiment.) We may now imagine the machine exposed to a Turing Test. The machine works in the following way:

The interrogator types in a conversational turn. The machine compares this with all the first turns of the strings in its memory. It selects from the memory the many, many conversation strings that begin with this turn. It picks one of these strings *at random* and replies with the conversational turn that happens to come next in that conversation string. The interrogator responds. The machine now compares these three turns with its memory, selecting the many conversation strings that begin in the same way. Again it selects one of these conversation strings at random and responds with the fourth turn in that particular string. Continuing in this way, at the end of an hour, the machine and the interrogator have reproduced one complete conversation from the memory store and, because all conversations in the store are reasonable, the computer's performance ought to be indistinguishable from that of a real human being such as the one sitting in the control cell.

Block's (1981, p. 22) description is apt. He says the machine acts as a conduit for the programmer's intelligence:

> If one is speaking to an intelligent person over a two-way radio, the radio will normally emit sensible replies to whatever one says. But the radio does not do this in virtue of a capacity to make sensible replies that *it* possesses. The two-way radio is like my machine in being a *conduit* for intelligence, but the two devices differ in that my machine has a crucial capacity that the two-way radio lacks. In my machine, no causal signals from the interroga-

tors reach those who think up the responses, but in the case of the two-way radio, the person who thinks up the responses has to hear the question. In the case of my machine, the causal efficacy of the programmers is limited to what they have stored in the machine before the interrogator begins.

Block (1981, p. 21) says of his machine that "All the intelligence it exhibits is that of its programmer." The intelligence is captured in the programmer's choosing the set of conversations that appear sensible to him and excluding the others.

The Machine's Strengths

In spite of its essential simplicity of design, the Block machine is very good indeed. It can answer any technical question that its programmer can answer and its interrogator knows how to ask. It can appreciate and tell jokes, and it can write or understand any poem that its human programmer can understand.[3] It can cope with all the tricky questions such as, "Do you love me?", "What does it feel like to hate?", "How does it feel to eat an onion?", "How did you learn to ride a bike?", and "Continue the sequence '2, 4, 6, 8'" in a variety of spoken contexts. It also could have just as good a conversation about chess as its programmer.[4] Insofar as a successful conversation machine needs to hypothesize about the internal states of its conversational partners, or build a model of its user, the Block machine will effectively do it. It will appear to grasp every nuance of every form of words and every implication of social context that its interrogators are able to express through the medium of the typed word at the time it was programmed.

The machine also would be a super expert system in any field of competence covered by the programmer so long as all that was required was written advice or instruction. Thinking about written conversation in this way helps resolve problems about the relationship between skill and what can be said about it. For instance, *in a Turing Test* the machine could pass as someone from Semipalatinsk so long as its programmer was a native, even though it is not socialized itself. A machine as big as the Block machine could be programmed with everything that can be typed about Semipalatinsk even though life is too short for the same to apply to a person.[5]

The same argument applies to a person skilled or experienced in an art or craft. A human pretending to be skilled in a Turing Test is likely to be caught out by an interrogator who is skilled. An inexperienced human will not know enough to generate the full range of relevant conversations. It is just a matter of logistics.[6] But a Block machine would not be caught out because it can be programmed with every conversation in which a skilled person could take part.

Socialization turns out to be important to the human spy in a Turing Test-like situation for logistical, rather than more fundamental, reasons. But in such circumstances the Turing Test is not a test of a skill but only a test of what you can say about a skill. Having a skill involves more than talking about it as I will go on to show.

The Machine's Weaknesses

Earlier, when I suggested that the programmer would not select all sensible strings because he might not recognize passages of Wittgenstein, A. A. Milne, or Garfinkel as sense, I glossed over a problem. The problem is that the programmer cannot anticipate what every interrogator might say. If an interrogator's conversational turn included something that the programmer thought was an impossible passage, the program would grind to a halt. There is, however, a technical solution to this problem. The programmer can be liberal with what he counts as potential interrogator-turns; he can include almost everything. To make the program work he will select strings that contain these turns followed by turns along the lines of, "Sorry, I don't know what you mean by that." It does not matter if the programmer includes interrogator-turns that no interrogator would ever actually use because they will simply never arise during a Turing Test. Strings, including those turns, will never be selected from memory during a run.

Now let us examine the programmer's task in more detail. Imagine the programmer sitting down with the long list of symbol-strings and reading each of them through. He selects those that do represent reasonable conversations and rejects those that do not. In making the decision his first concern is whether the responses that the machine will make represent what he could have said in response to the same putative interrogator-turns.

Suppose the programmer is reading through some strings that contain a single political comment and response. For example $string_i$ contains $interrogator\text{-}turn_n$, which reads, "I prefer conservative politicians." This is followed by $machine\text{-}turn_n$, which reads, "I prefer more progressive types." Because the programmer is himself a "leftish" type, there is no reason for this machine-turn to seem anything other than reasonable, and he will continue to check the rest of the string for sense. Now suppose he finishes with that string and some time later starts on another, $string_j$, which is identical to $string_i$ except that $machine\text{-}turn_n$ in this string reads, "I prefer left-wing politicians myself." The programmer's decision will be the same as before; this variation is equally consistent with what he might say in response to $interrogator\text{-}turn_n$. Indeed, there are a very large number of possible comments that express a politically leftish sentiment that could be substituted for the original $machine\text{-}turn_n$ without causing trouble and the set of symbol-strings contains conversations that include all of them.

Now suppose the programmer comes upon $string_i$*, which is identical to $string_i$ up to and including $interrogator\text{-}turn_n$ but $machine\text{-}turn_n$ reads, "Yes, me too. I'm glad to meet another conservative." This is not the sort of response that the programmer would normally make. Should he include it?

Now, the programmer is supposed to be including only what could count as sensible conversations in English. This cannot simply mean grammatically sensible conversations for there are many such conversations that do not make sense. Any conversation that contained too many statements along the lines of "green elephants sleep furiously" would give the game away. The only safe course of action is for the programmer to include just those replies that he would be likely to utter himself.[7] His first instinct, then, on encountering a politically unsympathetic machine-turn will be to reject the whole string. Call this the *restrictive strategy*. If he continues this way, the machine-turns, it might be thought, will represent his own conversational repertoire reflecting, among other things, his political personality.

The trouble is that the set of strings that remain will not really represent his conversational repertoire. There will be no strings at all expressing conservative sentiments yet the programmer will probably be able to think of occasions or contexts when he could make such a response. For example, if he had a whim to pretend to be a conservative, or if he felt there was danger in expressing left-wing sentiments, or if wanted to play devil's advocate in an argument. If he rejects all strings that contain "rightish" sentiments, then he will not produce a a machine that has his full conversational potential.

In a Turing Test the interrogator now has an advantage. If he bets that conversational partners who always produce woodenly predictable political responses are machines, he would be likely to win his bet in the long term. In this case the program has not mimicked the full range of human competences.

An alternative strategy for the programmer, the *liberal strategy*, is to include strings that contain politically conservative machine-turns. Now, for each grammatically accurate conversation string that contains a left-wing sentiment among the machine-turns there is a conversation string that contains a right-wing sentiment. The programmer might express his left-wing sentiments in many ways and when he plays devil's advocate, or whatever, he has just as many ways to express right-wing sentiments. Therefore, if he wants to express the full range of his conversational potential he must include just as many strings containing right-wing sentiments as strings containing left-wing sentiments. This follows solely from the flexibility of the English language and the rules of selection under which the programmer operates; he must include all conversations that make sense to him.[8]

Under this liberal strategy there are as many strings expressing right-wing sentiments as left-wing sentiments in the memory of the Block machine. When the machine is interrogated it selects all the strings that are consistent with the conversation that has taken place so far and then selects the string containing the next turn *at random*. This means it is just as likely to select a string containing a conservative response as a progressive one. Thus, even though the programmer was unambiguously a progressive thinker, the actual political preference expressed by the program would be unpredictable. The ratio of conservative to progressive responses within strings would be the same as the ratio of conservative to progressive comments in the English language rather than the same as the ratio in the conversational performance of the programmer. This ratio would hold at every machine-turn expressing a political sentiment in the entire string, for the programmer *might* have decided to play devil's advocate at any time.

Under the liberal strategy the interrogator who bets that the machine is the politically inconsistent respondent[9] will win the bet in the long term.[10] One way of expressing the problem is to say that the machine's memory contains the programmer's conversational capacity but not the programmer's conversational propensities. Capacity without propensity does not model a human being.

The same arguments apply to nonpolitical preferences and, indeed, to all cases where we can imagine that a conversational partner might decide to break the rules of normal conversational interchange. Thus, consider another form of rule-breaking action: "nonsense talk." Block says that some of the strings that represent possible conversations would include turns such as, "Let's talk nonsense for a while," with some nonsense turns following, but this gives rise to a variation of the problem we have just encountered. The set of speakable strings that include nonsense-containing turns is much larger than the set of strings that contains only what we normally think of as sense. This is because nonsense includes many more arrangements of the 100 available characters than standard English. The set of nonsense-containing strings is so much less limited than the sensible strings because nonsense can be so freely created; there are fewer cultural restrictions on what counts as nonsense because nonsense involves breaking normal cultural boundaries.

Now imagine a conversation between speakers '*A* and *B* that does contain a bit of nonsense and would therefore be included in the Block machine's memory under a liberal interpretation of what constitutes conversational competence:

1A: Hello, Jim, nice weather today.
1B: Needle niidle noo.
2A: You naughty boy.
2B: Aaaaaaaaaaaaeeeeeeeeeeeeoooooooooooogh!——That's better.

Though there are some restrictions on their form, 1B and 2B are but single examples of a large number of possible continuations to 1A.[11] For example, while retaining the general form and rhythm, 1B might read "Needle noddle noo," "Needle niddel noo," "Noddle niidel noo," "Needle niidle nog," "Noggle needle niing," and so on for a very long time indeed. Given a liberal strategy, all these continuations will have to be included in the Block machine's repertoire of continuations to 1A. This is because, if it is to be indistinguishable from a certain class of humans, it will have to use nonsense itself from time to time in a spontaneous way; it will have to be as potentially inventive in its use of spontaneous nonsense as a human.

Now let us imagine that the interrogator begins with 1A—"Hello, Jim, nice weather today." The Block machine will look through the large number of strings in its memory that start in this way and choose one with its associated second turn. It might find a conversation string with a sensible continuation such as, "Yes, not bad for the time of year," or "I got a bit wet on the way to the test, actually," or "Why are you calling me Jim?" But there will be a much larger set of strings with nonsensical continuations because of the freedom that nonsensical inclusions give to the way letters can be combined. Thus, when the machine selects a single string at random from the range of possibilities there is an overwhelming chance that it will come back with a nonsense continuation. The problem of nonsense turns out to be even worse than the problem of political sentiment because the ratio of nonsense turns to sense turns is much higher than 1:1, and therefore the number of *potential* conversations containing nonsense is much larger than those containing only sense. This is so despite the fact that actual written human conversation rarely contains nonsense because humans don't have a great propensity to use it. It is their capacity to use it that wrecks the Block machine.

To reiterate, although there will be no string in the machine's memory that contains more than the amount of nonsense than can be supported in a foreseeable conversation among humans, it is overwhelmingly likely that unobjectionable nonsense will figure early and often in the machine's conversational repertoire. Though the programmer's typical hour-long conversations contain bits of nonsense, say, only one time in 1,000, the machine would almost always talk some nonsense whatever is said to it. Again, this shows the interrogator the way to bet.[12]

These problems might look like mere technicalities, but they are really instances of the profound difficulties arising out of the attempt to replace a socialized sense of context, which tells one when and how often to break rules, with behavioral coordinates of action. Going back to Block's metaphor, the programmer is like a conduit. What he appears to convey to the program as he selects the sensible strings from the nonsense is his social competence in separating reasonable interchanges from unreasonable ones. But the promise is not fulfilled. The machine is unable to recognize the moment for the introduction of a piece of innovative social behavior.[13]

Another problem is that the machine can only be programmed with what counts as acceptable conversation at a particular historical moment. Thus, a programmer like me would include strings containing "Twas Brillig and the Slithey Toves did Gyre and Gimble in the Wabe" and some bits of Goon Show-type dialogue, but I could not include the works of the next Lewis Carroll or the next equivalent of The Goons. Suppose the next lot of Goons turned up shortly after the machine was programmed. A whole lot of new responses would suddenly become widely legitimate, but the machine would not know about them. Of course, all those possible future Goon-like and Lewis Carroll-like responses can be found among the Very Very Large Number of symbol-strings, and I have already shown that the programmer has no problem if he includes them all among the interrogator-turns. But, if the machine is to imitate a human, such responses would have to be included in the machine-turns too. This is a different sort of problem altogether because an illegitimate machine-turn will be executed if it is in the memory even though an illegitimate interrogator-turn will not. If the programmer cannot anticipate what might become legitimate speech in English at the time the machine is exposed to interrogation, he will not know what to include.

Imagine the impossible. Imagine the programmer is as much of a creative genius as Anthony Burgess and can foresee the possibility of a whole new style of dialogue of the sort that Burgess invented for his novel *A Clockwork Orange.* The trouble is that the language of *A Clockwork Orange* has not, in fact, caught on. Invention of new languages may be the prerogative of individual geniuses but legitimation of new languages is the prerogative of the cultural collectivity. Languages are the property of communities not individuals.

Because the programmer cannot anticipate the future of language, for the machine to work lastingly, he will have to be on hand all the time, continually updating and changing the memory store so that it corresponds to the social changes taking place around him. He will have to act as a permanent "socialization conduit." In that case, however, the machine can be dispensed with—the programmer might as well speak directly to the interrogator just as in the two-way radio to which Block compares his device. Of course, exactly the same is true of the control. If the control is locked away from society once the test has begun, he too will never learn the novelties of the changing social repertoire. But we are not interested in comparing machines with prisoners or social isolates, only with the most human-like of humans.

Skill, the Turing Test, and Interpretative Asymmetry

Nonsense is only one example of nonstandard English. The less standard the English, the greater are the creative possibilities, and the greater the problem for the Block machine. For example, the phonetic transcriptions of naturally occurring speech, beloved of conversational analysts, show just how much work we do in understanding what is said to us. As an unintended consequence the transcripts also show us just how well we can repair and make sense of unusual *printed* representations of the language. Take the following quotation from Lynch (1985, p. 220):

Th'thing is's thee ehm
This is garbaijhe.
Wehh I dunnuh
Nuhh doesn' look like vesiculs
Hhlooks more like a spine er s'm

Just as there were many possible versions of the nonsense phrase "needel niidel noo," there are many possible versions of this kind of phonetic representation. For example, the words in the second line of the transcript (I hypothesize) might on other occasions be better represented as "This is garbaijjhe" or "Thisis garbaijhe." At issue is not the accuracy of Lynch's transcripts but the flexibility of speech as opposed to regular written language.[14] This means that phonetic transcripts of speech are far more flexible than their standard written counterparts. If all these possibilities are included in the memory of the Block machine, then, in the counterpart of the nonsense problem, we would see the machine resorting to phonetic spelling with unwarranted frequency. Both this and the nonsense problem are symptoms of the inability of the machine to deal with our everyday creativity in the use of language.[15]

This argument leads us to a much more important generalization. As has been said above, the feature of the Turing Test that loads it in favor of the machine and allows a design like the Block machine to be a contender is the restricted channel of communication. Written communication can transmit only articulated knowledge—it can transmit only what can be said about a skill, not the skill itself. Therefore, the interrogator cannot test the scope of the machine's tacit knowledge in a direct way, he can only test if the machine can *say* everything that an expert could *say* about a skill. For example, although the machine could tell us all about the skills of golf, it cannot be asked to demonstrate them. So while the Turing Test is a good test of human-like capability insofar as it is a test of interactional competence, it is a poor test because interactions involving skills are not possible.[16] It is not possible to check to see if the machine can *do* things by virtue of its tacit knowledge, only whether it can articulate whatever its programmers can articulate by virtue of what they can do.

And yet there is one skill that we can test directly even in the restricted circumstances of the Turing Test; this is the skill of making written conversation. The nonsense problem and the phonetic spelling problem are examples of failures in conversational skill. The machine cannot, after all, handle language. In chapter 13 I noted that Colby arranged intermediaries between his program, PARRY, and the human interrogators to overcome just this problem. This, as we can now see, was not just a matter of convenience. For computers that interact with us through the written word (most of them), the language interface is the point at which either our skills or the computer's skills are required to digitize the world. If the computer is ever to become genuinely intelligent it will cease to require our help; it will learn to digitize the world for itself in the same way as we do. It will have to learn our inductive propensities. This new ability will become visible at its interface with us. That is the point at which it has the chance to display a skill.

The development of the printing press led to standardization of the written English language. In the Turing Test the restriction of the channel of communication normally has the effect of making all legitimate conversational turns expressible only in the subset of standard English that comprises English as "defined" by the conventions of printing. This is English in which the words have been digitized, and it is the digitization that makes it possible for us even to imagine that we could capture all possible hour-long conversations in a program.

Computerized dictionaries reveal the problem in miniature. Such a dictionary contains a digitized subset of the language. Offer the computer something outside the subset, such as "niddle," and it asks if you really mean something else. *You* then have to decide whether to include a strange and temporary usage in your *personal* dictionary. If you do not include it, the computer's dictionary will always query it every time it appears in a text.[17] If you do include it, the dictionary will never query it again even if you put it in by mistake. The dictionary cannot solve the problem of when to accept a creative usage. The Block machine is an attempt to overcome this problem by creating a vast dictionary of huge "words." Instead of words, the dictionary contains whole conversations. That is why the machine is so big.[18] In the end, however, nothing is solved.

I have suggested that we test machines such as the Block machine by looking for their ability to handle the equivalent of pre-printing-press English, including phonetic spelling and nonsense with all their implications of nonstandardized openness and creativity. To handle this, the computer will have to learn our inductive skills.

A general conclusion about Turing testing follows from the argument. As I remarked in chapter 13, language handling is not an esoteric skill, it is spread widely among the population at all ability levels. It is, for example, a skill that we regularly exercise when we treat the output of computers with respect as when we make allowances for the curious mistakes of ELIZA-like programs, or when we hear poor voice output as human speech, or read the limited vocabulary of ordinary package programs. It is our ready willingness to repair such deficiencies that allows current computers to work with us. It is the invisibility of this repair work—because of its pervasiveness even in ordinary human speech—that makes us able to mistake machines for humans. If, however, in conditions of doubt, or in Turing Test-like circumstances, we exploit interpretative asymmetry and instead of repairing the computer's awkward speech we look for its ability to repair ours, we will discover whether we are communicating with an entity that has skills. If we ever make a computer that can handle the full range of conversational skill, even via a keyboard, then it will merit the title "intelligent machine" or "artificial expert." Such a machine will have to share our culture.

V

Final Remarks

15

Intelligent Machines: An Experiment in Knowledge Science

I began with some major claims. I argued that intelligent computers could not be treated as isolated brains. Computers have to be thought of as social prostheses—replacements for humans in communities. Just as the potential of an artificial heart can only be understood in the context of the body, so the power of a computer can only be understood in the context of the social group to which it contributes. Knowledge science is the science of knowledgeable communities, and therefore understanding computers is a proper subject for knowledge science and artificial intelligence is an experiment within knowledge science. The argument applies just as much to other artifacts, including slide rules, logarithm tables, and printed books; in use, all have to interact with knowledge communities. To understand computers is to understand books and slide rules, whereas to understand books and slide rules is to understand computers. The question is, "Given that intelligent machines, like calculators, slide rules, logarithm tables, and books in general, are social isolates, how do they work? What happens to social groups when humans are replaced by things?"

Part of the answer is that the humans compensate for the deficiencies of artifacts in such a way that the social group continues to function as before. Of course, the group must have the expertise and the will to make-up the deficiencies. If we have a great deal of respect for our artifacts we will be so ready to help that we will not even notice their failures. In this way it is possible for a program like ELIZA to be used as a psychotherapist, and because of this it is possible to make easy use of pocket calculators, slide rules, logarithm tables, and books. Most of us are sufficiently expert in arithmetic and ordinary language to compensate easily for what these artifacts cannot do. But in a society without such expertise neither slide rules nor calculators can do arithmetic while books remain silent.

The contribution of the group is, however, only part of the story. My pocket calculator does do *something* that I do not do. If I take the battery out, it will cease to do it; I cannot fully compensate for the deficiencies of a broken calculator. What is the difference between a calculator with a battery and one without? What is it that comprises the calculator's abilities and how does it communicate with us?

The solution is very simple. It lies in the extraordinary abilities of human beings. Humans are so versatile that they can mimic artifacts. Wherever we choose to mimic a thing, a thing can mimic us. Mimicking an artifact is done by carrying out a special kind of act—a behavior-specific act, or machine-like act.

Existing critiques of intelligent machines divide the world of knowledge into two kinds: formal and informal knowledge stuff. This creates the appearance of a knowledge barrier that computers cannot penetrate. The knowledge barrier is spurious. Nevertheless, because computers can do certain things and cannot do others, the world of knowledge must be divided into two in one way or another. I replace the dichotomy of knowledge types with a dichotomy of human action—regular action and behavior-specific action. This explains what computers can do and what they cannot do, but it leaves room for them to creep across the boundary by incremental growth. Regular acts can be mimicked more closely by ramifying sets of rules and behaviors.

Ramification of rules and behaviors will never fully substitute for regular action though retrospective analysis disguises the difficulty. Failure to make progress is often blamed, not on the fundamental nature of knowledge, but on overlooked deficiencies in the rules. Programmers and knowledge engineers try to make rules reflect reality more and more accurately, but accuracy is not the essence of skillful performance. It is a mistake to try to counterfeit skill with *hyper-accuracy*.[1] Nevertheless, it is also a mistake to underestimate our ingenuity when it comes to mechanizing those elements of human action that can be satisfactorily executed in a machine-like way.[2]

Skill, Talk, and the Theory of Machine-like Action

Expert systems are machines that use what experts can *say* about their knowledge as the basis of their programs. It is easy to show that the whole of an expertise cannot be captured in what can be said about it. Asking, "How does an expert system fit into a social group?" is a more taxing question. The relationship between what people can do, what they are able to say about it, and what they choose to say about it varies from skill to skill and social location to social location. This affects the way knowledge acquisition should be approached. The fundamental inexplicability of knowledge only causes insoluble problems if it is expected that the system will be used by nonexperts. Otherwise, systems can be useful advisors to experts in rather the same way that pocket calculators are useful advisors to expert arithmeticians. (Calculator users are expert arithmeticians whether they realize it or not.)

One problem for knowledge acquisition is that some rules of thumb apply only to humans as physiological and psychological entities. Such *tangential* rules include, "Hum the 'Blue Danube' as you play a round of golf." More significantly, it is easy to read too much into the fact that humans often execute a skilled performance least clumsily when they do not think about it. Though there may be an important psychological or physiological difference between self-conscious and unselfconscious performance of a skill, the crucial philosophical distinction lies on an orthogonal dimension. The crucial distinction is between behavior-specific and nonbehavior-specific acts. Both types of act may be accomplished either self-consciously or unself-consciously. The possibility of mimicking an act mechanically depends not on whether the rules of performance are interpreted self-consciously, but on whether the act has the potential to be performed in a behavior-specific way. The argument can be set out schematically (figure 15.1).[3]

It is box 4 in this table that causes most confusion. This box represents two kinds of activity. There are activities like paint spraying, the rules of which have never been explicated. Nevertheless, so long as we are talking about that part of paint spraying that is carried out in a behavior-specific way, it can be mimicked without explicating the rules using record-and-playback or something similar. Box 4 also contains activities like military drill, the rules of which may be followed unselfconsciously by an expert or self-consciously by a novice (when box 2 is the correct location). Mimicking such a skill is particularly easy, but wherever the internalized act is machine-like, novices' rules are sufficient to mimic the act in its entirety.

Because it has sometimes been thought that all expertise is in principle inexplicable, too much may be read into the perform-

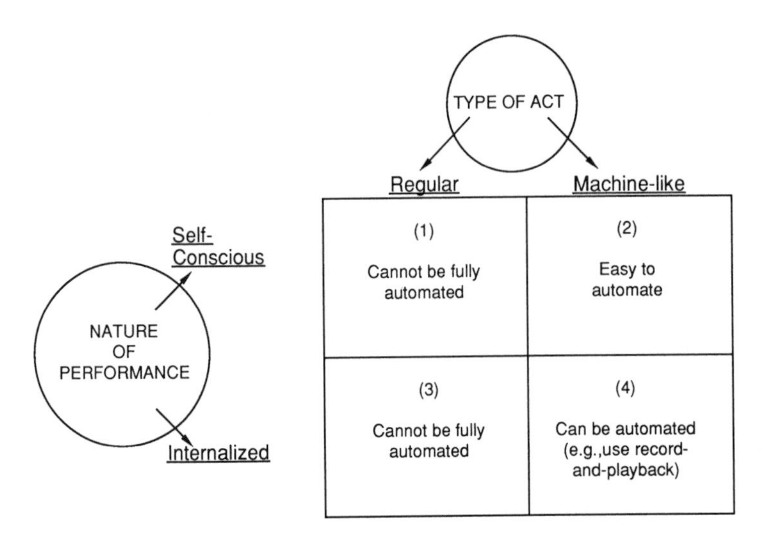

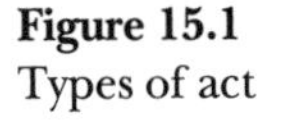
Figure 15.1
Types of act

ance of machines that do reproduce aspects of skill that are normally unselfconsciously practiced. For example, neural nets seem to have had some success with transforming text to speech although neither programmers nor speakers can explicate the rules of performance. Though the method of achieving this performance may be technically innovative, the crucial question is whether the device does more than can be described by machine-like rules. Just how the rules are encoded in hardware and software is not the crucial question.[4] To repeat, the crucial question is whether something found in boxes 1 or 3 has been accomplished, or are we still in the realms of boxes 2 and 4.[5]

Knowledge belonging to box 3 is sometimes referred to as "embodied knowledge." A good, literal example is the knowledge of a brilliant tennis player. If the tennis player's brain were transferred to someone else, we would not expect the recipient to be able to play as well as the donor because the recipient's muscles and body would not be developed in the right way. Although this example shows the fallacy of thinking about skill too cognitively, the principle of embodied knowledge applies equally to purely mental activities. Perhaps we should talk of "embrained" knowledge. The point is that just as the tennis player has a physical skill, the mathematician—and I do not refer here to the machine-like aspects of mathematics—has a skill that is embodied in the brain. It is not to be thought of in a cognitivist way just because the organ in which it is located is the brain rather than the body. Neither embrained nor embodied knowledge can be explicated; it cannot be taken out of the respective organs.

On the other hand, the knowledge that belongs to box 4, unthinkingly executed though it is, is neither embodied nor embrained. It is better thought of as *encoded* within the brain or body; a code can be broken and its contents can be read. The brain and body contain some types of knowledge in the same way as a book; it is just that the mechanisms for reading the knowledge are different in the case of book and brain or body. If the print is small, a book may need to be read with a magnifying glass; a microfiche needs a fiche reader; a floppy disk requires a computer; Morse Code needs decoding into a natural language. The machine-like knowledge encoded in a human being has to be read in a still more elaborate fashion: the human must be stimulated in various ways (such as shouting, "Right wheel!") and the responses noted. Once read in this way the knowledge can be reproduced in other forms, such as books or computer programs. Decoding only works, of course, where visible behavior is regularly related to stimuli—the case of behavior-specific action. In the case of ordinary action — where knowledge is embodied or embrained—rules cannot be decoded from behavior. This is the source of many mistakes in the social and behavioral sciences.[6]

Knowledge and Practice

The conventional status hierarchy of types of knowledge (figure 8.1) undervalues practical abilities. This causes difficulties when experts are asked about their practices. Experts prefer to think of knowledge in the well-ordered categories of the textbook rather than as inhering in the things they do every day. It is unreasonable to expect to find phenomenological self-consciousness in a craftsman who has grown up in this Western way of thinking. To notice what we do every day takes sociological or philosophical training. Textbook knowledge, on the other hand, is always unfamiliar and therefore always visible and ready to be expressed. No special effort of introspection is required to separate it from us.

Nothing can be described in full. In principle this whole book could be filled with a description of just half a minute of crystal growing experience and still not begin to exhaust it. Like silk scarves pulled from a bottomless conjurer's hat, the more one delves into knowledge, the more one finds. The job of the knowledge engineer is to take the ever-ramifying scarves of knowledge and cram them into a cloth cap.

To think this is easy is not only to misunderstand philosophy but also to misunderstand the physical world. Often, the mistake happens when well-ordered scientific descriptions of the world are taken to be the world itself. The knowledge contained in textbooks is easy to capture in an expert system, for in textbooks everything has already been parceled and tidied. Try to describe day-to-day practice and it bursts from the wrapping. Even the closed universe of a sealed ampule containing but three elementary substances, turns out in practice to be complex beyond prediction. This is not for any deep mathematical reason—for example, there is no reason to suppose it is a *chaotic* system—it is because there are too many variables to be measured. Knowledge that is high up in the conventional hierarchy is no use in such a case. Call this the *sociological uncertainty principle*. The principle states:

When a system is completely understood, it is too late for all practical purposes.

Where the principle applies, only knowledge from low down the hierarchy is useful. This knowledge is not found in texts or computers but only in skillful people.

The crystal-growing experiment shows why an expert system can never be an apprentice master. It shows what users contribute. One can see that CRYSTAL would work for our expert, Draper, because he could make the necessary contributions.[7] One can see equally why the system would be no use to you or me. Even if all the information in chapters 11 and 12 of this book were loaded into NEOCRYSTAL (it could be done now that we have the advantage of hindsight), NEOCRYSTAL would provide a recipe for growing just some crystals in the Bath laboratory.[8] We would not have avoided the situated particularity of crystal growing knowledge. To grow a new crystal for some new purpose we would still need Draper or another teacher to suggest or invent a method and, crucially, to be a channel from crystal-growing society to tell us what would count as a crystal in that particular instance.

What would it mean to solve this problem? What would it mean to have a system that required *no* help from its end user? What would it be to have a stand-alone expert system that contained all the power of an apprentice master? It is very hard even to think about the answer because such a machine would be like nothing on earth. It would not be like a human being for there are no stand-alone human beings. Human beings are not repositories of culture, they are reflections of it. They also maintain it and repair it. Appearances to the contrary, no one is apprenticed to one master because even a single master is but a participant in and a reflection of society. Through our masters, we are apprenticed to society.

Deskilling?

Can machines take over the skilled jobs of workers leaving the residue of the work force deskilled? Yes, but only for behavior-specific skills. Machines cannot replace ordinary skills. Insofar as there is a process of deskilling taking place on the factory floor it is not a matter of what machines can do but of what we can do. If we mimic machines in our actions, then machines can take over from us, but the job must already be machine-like before the machines are in place. Those who want to substitute human labor with machines must first arrange the job so that it can be done in a machine-like way; that is where so-called deskilling comes in.

An *ideal* Taylorist production line is as much a factory without ordinary skills as a factory without any humans at all. On the other hand, there are factories full of automatic machines where skill plays as great a part as ever it did. For example, Jones (1989) explains that under some regimes the shop-floor operative still has the responsibility of adjusting the program of automatic metal-cutting machines. Skills are just as much in evidence, they are simply channeled through the machine in a different way.

Of course, even where there is a deliberate regime of deskilling, residual skills remain. Harper (1987) describes the changing life of an isolated mechanic skilled in the repair of Saab motor cars. He sees the mechanic's skills becoming less necessary as larger and larger components are treated as sealed units. He refers to modern mechanics as mere "parts exchangers." And yet those who have ever tried to exchange a part on a car will know that the skill of loosening a rusted nut (graphically described by Harper), is unreduced. How hard should one hammer or wrench before something breaks? Changing a part is not a negligible skill.[9]

My word processor helps with the mechanical parts of book writing. Machines, intelligent or otherwise, increase the efficiency of factories and individuals. Tools make jobs easier. Some tools, like refrigerators and space rockets, make it possible to do jobs that no one could do before. Intelligent machines are useful as tools. The only mistake is to think that a tool is an actor. It is easy to say, "A calculator can do arithmetic better than I." But the epistemological sense of that statement is identical to the sense of, "A hammer can drive in nails better than I." To mistake the calculator for the arithmetician is like mistaking the hammer for the hammerer.

Turing's Sociological Prediction

Consider once more Turing's prediction. By the end of the century he expected that

> the use of words and general educated opinion will have altered so much that one will be able to speak of machines thinking without expecting to be contradicted. (Turing 1982, p. 57)

There are at least four ways in which we might move toward such a state of affairs: (i) Machines get better at mimicking us; (ii) We become more charitable to machines; (iii) We start to behave more like machines ourselves; (iv) Our image of ourselves becomes more like our image of machines. Let us consider each of these in turn.

(i) Machines get better at mimicking us

Unless machines can become members of our society, something that is unforeseeable, they can appear to mimic our acts only by developing more and more ramified behaviors. This process is a good thing so long as it is not misconstrued. Ramification of behaviors makes for new and better tools, not new and better people. The rate at which intelligent tools can be improved is not easy to predict. The problem is analogous to predicting the likelihood of the existence of life on other planets. There are a large number of other planets but the probability that the conditions for life exist on a planet is astronomically small. Where two large numbers have to be compared a very small error can make a lot of difference to the outcome. The two large numbers in the case of intelligent machines are the exponential growth in the power of machines and the exponential increase in the number of rules that are needed to mimic action as the users of machines make less and less contribution to their performance. My guess is that progress is slowing down fast, but the model sets no limit to asymptotic progress.

(ii) We become more charitable to machines

As we become more familiar with machines we repair their deficiencies without noticing. This is what has led to us to speak of calculators as being so good at arithmetic. I imagine that those who first built calculators did notice their deficiencies, but *the use of words and general educated opinion* has changed sufficiently to allow us to talk of calculators in the fashion that Turing predicted would apply to intelligence in general. This makes for easy use of calculators but makes us lose sight of our contribution. To generalize this process is to lose sight of all our special abilities.

(iii) We start to behave more like machines ourselves

Consider the deficiencies of machine translation. One solution is to standardize the way English is written (Price 1989):

> In the US, it is usual practice amongst some large firms to send entire manuals for online translation on a mainframe computer. It works well. The manual writers are trained to use short sentences, cut out all ambiguities from the text (by repeating nouns instead of using pronouns for example) and to use a limited, powerful vocabulary.
>
> Europeans have shown little of this discipline.

For these "Europeans" there are, of course, no significant ambiguities in the texts they write, it is just that the texts are not written in a behavior-specific way. In what is usually counted as good writing, different words are used to represent the same idea and different ideas are represented by the same words. The parallel with regular action is complete. The problem is that translation machines—string searchers apart (chapter 14)—cannot participate in acts of writing unless they are machine-like acts. Only machine-like writing, in which the same word behaviors correspond to the same ideas every time, can be completely translated by machines. If we adjust our writing style to make it universally behavior specific, then mechanized translators will be as good as human translators, and we are more likely to come to speak of machines thinking just as Turing predicted. What is true of translation is true of all our actions. But we do not want to lose our freedom of *action* to accommodate machines.

(iv) Our image of ourselves becomes more like our image of machines

If we think of ourselves as machines, we will see our departures from the machine-like ideal as a matter of human frailty rather than human creativity. I have tried to counter the tendency to think of humans as inefficient machines. I have argued that there is an essential difference between the way humans act and the way machines mimic those acts. I have tried to show that machines can mimic us only in those cases where we prefer to do things in a machine-like way. Whether we come to *speak of machines thinking without fear of contradiction* will have something to do with whether my argument is more or less convincing than the arguments of those who think of social life as continuous with the world of things. My argument began with the contradiction between what computers had achieved and the social view of knowledge. The theory of machine-like action resolves the contradiction. So far, the experiment in knowledge science that is artificial intelligence points unerringly to the irreducibly social nature of human beings and their knowledge.

Intelligent machines are among the most useful and interesting tools that we have developed. But if we use them with too much uncritical charity, or if we start to think of ourselves as machines, or model our behavior on their behavior, not only will we lose sight of what we are, the tools we make will not be good ones.

Notes

Chapter 1

1. In this, knowledge science is quite different from cognitive science.

2. The natural sciences have been the first field site for this kind of study, giving rise to the modern-day subject "history and sociology of scientific knowledge." Knowledge science is more general in scope.

3. Indeed, the link is made quite explicitly in work that attempts to computerize scientific discovery (E. G. Langley et al. 1987). I will discuss such programs in chapter 5.

4. Though the recent studies are often thought of as post-Kuhnian, Kuhn's (1962) book, along with other philosophical writing such as Feyerabend (1975), provided a more philosophically liberated atmosphere in which the studies could take place while the significant technical input came from elsewhere. Perhaps the most important input was Wittgenstein's later writings (e.g., 1953). For the significance of Wittgenstein for science and mathematics see Bloor (1973, 1983). Equally influential for another group of writers (e.g., Lynch 1985) was the ethnomethodological approach of Garfinkel (e.g., 1967). Phenomenology (e.g., Schutz 1964) provided another source, as did the anthropological and semiotic traditions, notably in the work of Latour (e.g., 1987).

Actually the first case study to be done in the tradition of "Sociology of Scientific Knowledge," though it was not recognized as such, was done in the 1930s by Ludwik Fleck (1979). Fleck described his own work on the Wasserman reaction for syphilis. Fleck's work was little known until its much later publication in English. Other early case studies include Collins (1975) on Joseph Weber's nondiscovery of gravitational radiation and Latour and Woolgar (1979) on the discovery of a brain peptide. Detailed empirical studies flourished in the early 1980s. Collins (1981a) is an edited collection of original pieces that look at controversies concerning the discovery of magnetic monopoles (Pickering 1981), nonlocality in quantum theory (Harvey 1981), solar-neutrino detection (Pinch 1981), chemical transfer of learned behavior in worms and rats (Travis 1981), and the next stage of the gravity wave story (Collins 1981b). A slightly different approach may be found in Knorr-Cetina's (1981) study of a food research laboratory. Some of the interesting papers are collected together in Barnes and Edge (1982) and Collins (1982), while Collins (1985) brings together my own work on gravity waves, the TEA-laser, and aspects of parapsychology in an overarching framework. Pinch (1986) is a book-length study of solar-neutrino detection. Recent notable historical treatments using some of the ideas of this tradition of work are Rudwick (1985), which looks at a controversy in geology, and Shapin and Schaffer (1985), which concerns Robert Boyle's work on vacuum pumps and his controversy with Thomas Hobbes.

There are many more case studies that can be discovered by following the bibliographies in the works mentioned above. Recently, various schools have adopted different approaches based on exploration of the methodological implications of sociology of scientific knowledge. These, however, are not of great importance for this book.

5. To see just how much can be said about Semipalatinsk in a Turing Test-like situation, see chapter 12.

The spy is being taught on a kind of behaviorist model where all the inputs and outputs are immediately visible and explicit. Output is either corrected or reinforced. The problem is the same as is encountered with the recently fashionable parallel distributed processing (McClelland et al. 1986). The idea is attractive because it requires no explicit programming, but the inputs and outputs must still be controlled, and must still follow the inadequate behaviorist model (Papert 1988). Typical problems of instability will arise (Dreyfus and Dreyfus 1988) just as in machine induction for expert systems (See chapter 8).

6. In the absence of an extraneous disturbance the spy might claim some disability in his Semipalatinsk-ness—perhaps forgetfulness or an unusually short period of socialization or an old war wound to the head. An incompetent Semipalatinskian may do no better than the spy, it is true, but we are not concerned with comparing machines with incompetents—that is too easy for the machine (see chapter 11).

7. Lenat's ambitious project (Lenat et al. 1986) for encoding the common sense required to use an encyclopedia into an expert system can be likened to the problem of trying to train the spy.

During the industrial revolution, those who wished to steal another country's secrets had not only to acquire the new machinery, but also to suborn the operatives, see, for example, Jeremy (1981). For early coal-mining skills see Harris (1976).

8. I am grateful to Don McCloskey for drawing this passage to my attention.

9. I am grateful to Susan Newman and Jeff Schrager for pressing this point home.

10. Of course, if this were a possibility then we must already know about everything that needs describing in order to describe ourselves as physical beings—a conceit of the age. Even the last quantum state may not be good enough, however. Modern physics is just one abstract description among potentially many.

11. There is an argument about what is meant by the workings of the brain. Some proponents think that thought comprises symbol processing so that the mechanism by which the inputs and outputs are produced does not matter so long as these are identical to those produced by the original. On the other hand, Searle (e.g., 1984) argues that there is more to thinking than processing symbols and that the mechanism itself is the cause of consciousness, understanding, or whatever.

12. Ernest Gellner (1974) coined the term charitability. Gellner was complaining that anthropologists are too charitable, managing to interpret anything a primitive people did as rational within their own terms and thus leaving no room for any criticism. For a useful collection of readings on the "rationality debate," see Wilson (1970).

Chapter 2

1. There has been an argument about whether Dreyfus correctly predicted the progress of chess-playing computers. It seems to me that this argument turns on a willful misunderstanding of the main thrust of his analysis. For a résumé of the debate, see Turkle (1984).

2. For an extended treatment of the Wittgensteinian argument in the context of the replication of scientific experiments, see Collins (1985).

3. The argument about rules disposes of cognitivism only.

4. In this they are typical of those brought up in the modern Western scientific tradition, which takes propositional knowledge as the paradigm case of what it is to know something.

5. I leave aside moral capacity and emotional sensitivity, which are not the subject of this book.

6. For recent works showing that science is less formal than it conventionally appears, see chapter 1, note 4. For mathematics specifically see, for example, Wittgenstein (1956), Bloor (1973, 1976), and Lakatos (1976).

7. Bearing in mind that Dreyfus's predictions, and those which grow out of related critiques, have proved so accurate and lasting that any replacement must subsume them rather than meet them head on.

8. Note that the material that carries the symbol is almost irrelevant. That is why digital systems can be instantiated on any kind of hardware. Applied to artificial intelligence this is cognitivism.

9. Though if we were to try to settle an argument about relative value by weighing our pieces of gold, we would represent their weights by a set of digital symbols; every weighing machine has a finite number of columns of digits, and each column contains only one of ten symbols.

10. Digitization solves Plato's problem, as one might say. It divides up the world into discrete idealized entities. In such a world one may have essential tables and essential chairs because there is nothing in between.

11. I am grateful to Richard Southal and Debra Adams for long conversations about typesetting. It was impossible not to be infected by their enthusiasm for the subject.

12. I owe the joke to an anonymous reviewer.

13. Part of this commitment to reading symbols in only one way can be understood in terms of the notion of "incorrigibility" (e.g., Pollner 1974). That is what I have called, in chapter 1, the routine maintenance of ways of going on. Incorrigibility does not capture the much cruder, behavioristic aspect of the force of symbols that emerges from repetitive training and which I discuss in the next chapter.

Chapter 3

1. The larger part of Winograd and Flores's (1986) critique turns on a subtle point drawn from Searle's work on the idea of a speech act. The critique is not central to this book but is interesting enough to be worth noting. They argue that a machine does not have any moral commitments. Therefore, a machine cannot

engage in speech acts because speech acts require moral commitments. For example, when a human provides even a factual answer to a question there is an implication of moral commitment to truth. This is what puts the listeners in a position to act on the information. Social life is a web of moral commitments. The trouble with Winograd and Flores's critique is that it ought to apply just as much to calculators and books, yet such artifacts do share our social life quite successfully.

2. Though in this chapter I talk only of physical behavior, I intend to draw a parallel between this and what one might call *mental acts* and *mental behavior.* In discussing artificial intelligence, mental behavior is the crucial category. Nevertheless, it is much easier to begin the argument and to set up the major distinctions with concrete examples drawn from the area of actions with physical counterparts. In the end, separating acts that have physical outcomes, such as raising an arm, or going for a walk, from acts which we think of as happening mainly in our heads, such as multiplying 69 by 2.54, is not useful for our purposes. Attempts to use computers to mimic human intelligence are similar in most significant ways to attempts to mimic physical action with a physiological theory or mechanical simulation, so I will finish by making no significant distinctions between cognitive and physical acts.

To anticipate the argument and to ease the way into treating the cognitive and the physical sphere as indistinct, think about mechanical devices that accomplish cognitive tasks. For example, think about the abacus or slide rule, which I will deal with at greater length, or the computer language PROLOG. These are mechanical ways of accomplishing logic and arithmetic and can be thought of as related to logical and arithmetical *acts* in the same way as physical behavior is related to physical acts.

3. Of course, the "in principle" question is of interest to psychologists for it shows that to mimic the inputs and outputs of a human does not necessarily lead to more understanding of what goes on inside the head.

4. Baker and Hacker (1985, p. 166) go on to add:

> What counts as doing the same within a practice is determined from the perspective of the practice itself and is not responsible to an external reality

Searle (1984, pp. 57–58) puts it this way:

> At first it is tempting to thank that types of actions . . . can be identified with types of bodily movements. But that is obviously wrong. For example, one and the same set of human bodily movements might constitute a dance, or signaling, or exercising, or testing one's muscles, or none of the above. Furthermore, just as one and the same types of physical movements can constitute completely different kinds of actions, so one type of action can be performed by a vastly different number of types of physical movements. Think, for example, of sending a message to a friend. You could write it out on a sheet of paper. You could type it. You could send it by messenger or by telegram. Or you could speak it over the telephone. And indeed, each of these ways of sending the same message could be accomplished in a variety of physical movements. You could write the note with your left hand or your right hand, with your toes, or even by holding the pencil between your teeth.

5. Rose (1988, pp. 30–31), describes Taylor's ideas as follows:

> Pick ten to fifteen men already skilled in the work whose "science" is to be discovered; observe them at work and define the elements of the sequence of operations they employ; time each element for each individual with a stop-watch; identify the operations that seem to contribute nothing towards the completion of the task and eliminate them; select the quickest methods discovered for each element and fit them into sequence, forbidding any deviation or the introduction of unnecessary operations; . . . The result will be the "quickest and best" method

for the task. Because it is the "best way," all workers selected to perform the task must adopt it and meet the time allowed. Training a first-class man involved drilling into him a perfect grasp of the routinized movement of a given task. . . . Refresher courses should be provided to eradicate vetoed methods that crept back in, and it was imperative to convince the worker that he could not know better than the scientific manager how a job should be done.

6. Of course, in the case of Taylorism the manager collaborates in treating the worker as a feature of the natural world. In that case the theory also has a moral dimension.

7. Kusterer's (1978) book *Know-How on the Job* provides a number of examples of the use of skills in the lower levels of the workplace. Also see Jones (1984) and Jones and Wood (1985).

8. In practice it might be more efficient to go for recognizing chair parts rather than whole types of chair, but the principle remains the same.

9. Again, for what is almost a caricature of the method, think about army technical training. "The naming of parts" and the dismantling and assembly of guns, and so forth.

10. Others might prefer to skip the next few paragraphs and move on to the next section.

11. And, more contentiously, animals and young babies.

12. Of course, what scientists see as constant within the natural world changes as our scientific worldview changes. Nevertheless, our idea of the ultimate constancy of the natural world remains so long as science remains (see Collins 1989, 1985).

13. Not that this is the only type of machine. It would be interesting to develop a classification of machines based on their relationship to human competences. For example, the washing machine is some way from the paint sprayer in that it does the job reasonably with a rather different set of coordinates of action than is used by people. In general, it does its job less thoroughly than can be done by hand, but it does it well enough to be acceptable. There is no washing machine that can mimic human washing. The latter is too complicated and situationally responsive. Machines, such as refrigerators, do something for which there are no corresponding human coordinates of action at all.

14. For a similar argument in the context of rules of scientific method, such as "the rule of replication," see Collins (1989).

15. It is not hard to see how plans relate to closed systems and to microworlds such as that of Winograd's SHRDLU. It would be quite reasonable to expect plans to work in microworlds or in circumscribed domains, because nothing unforeseeable is going to happen. Particulars that might make real-world situations significantly different from one another are stripped of significance in a microworld. Everything in a microworld must be assimilated into some predefined class, such as a pyramid, a cube, something red, or something green. There cannot be anything that's not quite a pyramid in SHRDLU's world, even though our world is full of things that are not quite what they ought to be in the world of Euclid. What happens to the real world in a microworld is a process of idealization or digitization.

16. Apart from the general point being made, this problem is encountered in unmistakably concrete and particular form in "machine induction" and other attempts to mimic a naive model of scientific inference (see later chapters).

17. In a sense they are right. Without treads the wheels of the truck will spin in the sand and no progress will be made; without logic we cannot imagine progress in our reasoning. This kind of study of the fine structure of reason, however, explains nothing prior to the event (see Collins 1989).

18. In science studies it is the sociologists of scientific knowledge (e.g., Collins 1985) who have been stressing the gang-of-cutthroats model in the face of the rationalist philosophers' appeal to the orderliness or logic of scientific history and progress.

Chapter 4

1. The only models we have of learning are mechanical models, but these no more account for children learning language than they provide for computerized pattern recognition. For example, how do the children know when the sound that corresponds to a number begins and ends? The problem is similar to that of the recognition of the ambiguous photograph discussed in chapter 1.

2. Rote: mechanical memory, repetition, or performance without regard to meaning (*Chambers Twentieth Century Dictionary*).

3. Here I disagree with Dreyfus and Dreyfus's (1986) global account of the acquisition of skills as I will explain in chapter 6.

4. This is a matter of *response*, not recognition or interpretation, which are much more complicated. I will explain the distinctions in more detail later in this chapter.

5. For an account of learning to play the piano, see Sudnow (1978). This is discussed further in chapter 6.

Suchman (1987, p 19) captures the relatively poverty-stricken aspect of behaviorally specific instruction:

> A consequence of the human coach's method is that his or her skills must be deployed anew each time. An instruction manual, in contrast, has the advantage of being durable, reusable, and replicable. In part, the strength of written text is that, in direct contrast to the pointed commentary of the coach, text allows the *disassociation* of the occasion of its use. For the same reason, however, text affords relatively poor resources for recipient design.

6. I adopt here the notion of primary sameness discussed in chapter 3.

7. In writing these opening chapters I have been influenced by David Bloor, especially his comments on ethnomethodology. Bloor (1987) makes it clear that there is more to symbols than interpretation upon interpretation. His solution is to say that at some point we must "follow rules blindly." I cannot fault his philosophical argument, but I think Bloor associates following rules blindly with repetitive training of a sort that leads to the unthinking response to a rule that goes with internalization. I don't think internalization is the solution because on my theory there is no philosophical distinction between machine-like action done consciously and machine-like action done unthinkingly. I do not have a solution to the problem of the endless regress of interpretation that Bloor points to, but I do not think that the whole solution is to be found in behaviorist models of learning.

8. In some versions of relativism rocks too are treated in this way.

9. The problem is familiar to those who try to develop computer programs for reading. Though computers are completely responsive to our will—we can successfully impose upon them any program of training that we can think up—no existing computer can recognize varied typefaces with the same accuracy as a moderately literate human, and no computer is close to reading handwriting.

10. I am grateful to an anonymous reader of my manuscript for the following comment.

In *Drawing on the Right Side of the Brain* (NY: St. Martins, 1979 and later editions), Betty Edwards advises people to forget about the digitized, symbolic meaning of a house or a nose and just draw the shape. She advises us in fact to draw the shape made by the space around the shape, not [say] the nose itself directly. One of her exercises is to invert Picasso's famous line-drawn portrait of Stravinsky, so that we do not say "collar" to ourselves when drawing the collar-part. . . . "[T]ry not to think about what the forms are and avoid any attempt to recognize or name the various parts" [p. 53]. It works. When one reinverts one's own drawing it is startlingly similar to Picasso's.

11. The problem of symbol recognition is solved by making us digitize our input. That is why computers have keyboards. It limits our choice of input to about 100 possibilities.

12. I would guess that most of us first learned to do arithmetic by rote, without understanding the significance of, say, carrying numbers between columns. This, of course, is hardly surprising. There is no need to know what you are doing if it can be done by machine-like acts. Because the behavioral coordinates of machine-like acts can replace the acts without loss, learning to do the trick without understanding what is going on is enough for the purposes of passing examinations.

Jean Lave (1988) has found that pupils who can't or won't learn arithmetic at school—in terms of my account, they have not been prepared to surrender their freedom to a sufficient extent—may prove to be good at solving problems in real life. Lave refers to classroom arithmetic as an "esoteric" skill designed only for the educational world, whereas in practice arithmeticians such as weight watchers or supermarket shoppers find a variety of practical devices for solving problems.

One striking example is the use of the world itself as a calculating device. For example, set the problem of separating one-quarter by weight of a portion of cottage cheese, a weight watcher spread the cheese into a flat disk on a chopping board and divided it into geometrical quarters by eye. No conventional "sums" are needed to accomplish this.

On my account, what Lave has discovered are people who do their arithmetic in a much less machine-like way than we are taught in school.

13. See also Kuhn (1961).

14. We cannot delegate just anything. For example, we do not, as Latour (forthcoming) would say, "delegate" our journeys to our car. The car is not making a journey, merely following the behavioral coordinates of a journey (which in this case happen to be space-time coordinates). We can use the car to fit into our actions. We can save ourselves the trouble of the rote aspects of travel, and concentrate better on the act of traveling—a matter of going somewhere for some purpose. Thus, to adapt Searle's (1984) example, the act of walking south toward

Hyde Park is behaviorally indistinguishable from the act of walking south toward Patagonia. If we wanted to ride in a car, the car would begin the journey in the same way whether it was Hyde Park or Patagonia that we had in mind. Nevertheless, the acts are quite different, even though the car's behavior, and indeed our own journey described in terms of space-time coordinates, is the same (Searle 1984, p.58).

15. We an also see that a Chinese room might well be successful in a Turing Test if set to compete against someone who knew the Chinese language, but only if that person had learned the language in a mechanical way.

16. Of course, it is only concerted some of the time. We spend an awful lot of time arguing about it too. But without the underlying concertedness there would be nothing worth fighting for.

17. For instance, in the case of PDP (neural net) machines, think about how the training programs are used to constrain inputs into an exhaustive and previously defined set of possibilities.

Chapter 5

1. This is a less trivial matter than it seems. *The Experimenter's Regress* (Collins 1985) shows that in contested science knowing a result is not separable from discovering a result. This has significant consequences for philosophy of science. The paradox reappears in a different form where I discuss "weighing out" in chapter 11.

2. They are not only invisible to most of us, but they are scorned and undervalued even when they are described.It is, however, far from clear how we do these little things and, as for making computers that can manage them, the possibility is beyond the horizon.

3. Quite a lot has been made of the fact that some parallel-distributed processing machines make the same sort of errors in developing English reading skills as children. I would be most surprised if their comparative ability to cope with complexity and contextual dependency were like that of children.

4. The conventional base of simple mathematical series is explained in more detail elsewhere (e.g., Collins 1985).

5. Remember, too, that if you make a mistake your partner, recognizing that your intention is to execute the digitized movements, will be tolerant enough to repair your errors.

6. Actually, the machine would probably do a better job as far as the mechanical side of things is concerned.

Chapter 6

1. I was surprised to discover that an expert system for the domain of "love, intimacy, and friendship" has been written in all seriousness. See Fulda (1988).

2. McDermott (1981), in a very interesting article, cautions against the facile use of terms such as "learning" when applied to computers. He points out that anthropomorphic interpretations set up expectations in the minds of users that are bound to be disappointed. Expert systems do not really learn—they acquire more information.

3. Early attempts to replace human judgment with probabilities have not been a success. Human experts just do not think in these terms. I will not discuss probabilistic inferencing further.

4. The usage of the term "expert system" varies widely. For example, there are so-called expert systems that monitor information directly and respond automatically. Thus, there is a system that monitors the vital signs of hospital patients and adjusts conditions (or sounds an alarm) when danger appears. Similarly, expert systems are used to monitor and control manufacturing processes without intermediate human intervention. This kind of application is hard to separate from regular computer data handling; one would not think of subjecting such a system to a Turing Test!

5. For an optimistic forecast of the use of expert systems in this way see, for example, "Apprentices Lose Out to Computers," *New Scientist,* September 26, 1985, p. 34.

6. At stage two, the "advanced beginner" starts to take account of features of context that cannot in themselves be verbalized. For example, the advanced beginning driver may note the sound of the engine in deciding when to change gear. Such "situational" elements are mastered through experience and recognition of similarity to previous instances.

At Dreyfus and Dreyfus's (1986) third stage, "competence," the number of "recognizable context-free and situational elements" becomes overwhelming, and expertise becomes much more intuitive rather than calculating. "Problem solving" is no longer the predominant motif. The fourth stage in the account is "proficiency." The proficient expert recognizes whole problem situations "holistically" in the same way as the advanced beginner recognizes specific features of the environment. In the advanced beginner stage it is, say, the sound of the engine that is recognized from experience; in the proficient stage it would be a complete traffic scenario. Nevertheless, some elements of conscious choice and analysis remain to guide the proficient expert's decisions.

7. For pedestrians, one may think of the way we learn to cross the road. Explicit rules—"look left, look right, look left again and if nothing is coming walk rapidly across"—disappear as they become absorbed into the generalized skill of crossing roads. This is a skill that suddenly reappears, and has to be relearned starting with a conscious routine, when we go to a country where they drive on the other side of the road. Knowing how to cross the road is known by the novice as a set of explicit and fixed rules, but by the experienced road crosser as an unexplicated skill that is acted out in different ways as each new and unanticipated circumstance is encountered. As ability to cross the road increases, the pedestrian seems to *know* less and less about it. Actually, the experienced road crosser uses a nonmachine-like set of procedures having to do with making eye contact with the driver, and so forth. Learning to cross the road fits the Dreyfus and Dreyfus model quite well.

8. Thus, Dreyfus and Dreyfus deduce from their model that no expert system will ever surpass stage three (competence) because this is the best that can be done through recognizing context-free elements. (Though, as the number of elements requiring recognition is "overwhelming" even at stage three, their model might better restrict achievement to stage two.)

9. They have no satisfactory model of how past situations are recognized as *the same* as what is currently being encountered.

10. See Collins (1985) for an analysis of novelty in scientific research.

11. My thanks to David Edge for this clipping.

12. It is tempting to think that a work to rule is a matter of "literal" interpretation of rules.There can be no literal interpretation of rules , however, because rules do not carry with them a definitive interpretation. Working to rule, therefore involves a pedantic, rather than a literal interpretation.

13. This is actually a fictional example taken from an episode of a BBC TV series set in a comprehensive school, "Grange Hill."

14. Leith (1986) argues the importance of the continual development and establishment of new interpretations of rules in his critique of legal expert systems. It is not just a matter of precedent, but of arguments about what is to count as a precedent. The law par excellence is not a fixed body of rules but a set of moving precedents that rest on legal and moral authority.

15. It is not just a matter of reasonable response to novelty. Experts have the moral authority to establish proper usage as well as the ability to execute it. This point reflects Winograd and Flores's (1986) concern with the moral authority of speech acts.

Oddly enough, earlier on the same day I watched an equally extraordinary event on TV concerning a different sport. (Non-Commonwealth readers will have difficulty understanding the following account of an incident in a cricket match but might like to reflect on it as an illustration of the social location of language. Cricket is notoriously difficult to explain.)

In the Test Match between England and India, the Indian twelfth man remained on the field by accident while an English player bowled a complete over. What would have happened if a batsman had been out during this over? One commentator suggested that a batsman could not be out under these circumstances because the ball was dead during the entire over; the game of cricket is played between two sides of eleven men. On this interpretation, however, what ought to happen to any runs that England scored during the over? If the ball was dead, then England should not have been allowed to score—surely unfair if they had a successful over! Also on this interpretation, there is the question of whether an over was actually bowled—or did the same bowler effectively bowl twelve balls from the same end because the intervening over did not count? Again, if the over did not count as a proper over, did India bowl the number of overs required of them in the day?

The balls could not have been treated as "no-balls," or the over should have been six balls longer and England should have been given six extra runs. In any case, a run out, which is allowed off a no-ball, would not have been just under these circumstances because there were too many fielders. It is not at all clear what should have happened in this circumstance—certainly, the rules do not provide for it. In the event, everyone just forgot about it and got on with the game, reinforcing the rule that the "umpire's decision is final" if not setting any new precedents.

16. See Lave (1988) for the example of supermarket arithmeticians who do not make use of the coaching rules that the rest of us use.

17. This point is also to be found in the psychological literature. Reber et al. (1980) write, "complex structures such as those underlying language, socialization and

sophisticated games are acquired implicitly and unconsciously." (Quoted in Berry 1987).

18. This quotation is from a talk to The British Association for the Advancement of Science (Crothers 1987).

19. The Dreyfus and Dreyfus model captures only a part of what is going on.

20. Something similar can be found in the first paragraph of this quotation. In that paragraph the writer talks of Woosnam hammering in a three-footer. This is a most peculiar metaphor because it refers to rolling a golf ball about a meter over smooth, level ground into a small hole. The sort of violence normally associated with a hammer blow would be disastrous. In fact the metaphor refers indirectly to Woosnam's new-found confidence in his abilities, which enabled him to roll the ball firmly toward the back of the hole rather than trickle it along the ground in the hope that it would topple in.

21. For example, in writing this book I learned a great deal from the explicit hints of the readers of the manuscript.

22. Buchanan et al. (1983) make similar points:

> Knowledge acquisition is a bottleneck in the construction of experts systems. The knowledge engineer's job is to act as a go-between to help an expert build a system. Since the knowledge engineer has far less knowledge of the domain that the expert, however, communication problems impede the process of transferring expertise into a program. The vocabulary initially used by the expert to talk about the domain with a novice is often inadequate from problem-solving; thus the knowledge engineer and expert must work together to extend and refine it. One of the most difficult aspects of the knowledge engineer's task is helping the expert to structure the domain knowledge, to identify and formalize the domain concepts. (p. 129)

and

> Another major difficulty in knowledge acquisition is verbalization by the expert. It is almost always difficult for the human expert to describe knowledge in terms that are precise, complete, and consistent enough for use in a computer program. This difficulty stems from the inherent nature of the knowledge that constitutes human expertise: it is often subconscious and may be approximate, incomplete and inconsistent. (p. 154)

Interestingly, Hayes-Roth (1985) suggests that a useful rule for expert system builders is to "seek problems that experts can solve via telephone communication." Given the analysis here, one can see why this is such an apt rule—it ensures that the solution to the problem can be put into words.

For scientific attempts to improve knowledge elicitation and overcome the problem of tacit knowledge see e.g., Waterman and Newell (1971); Olson and Rueter (1987).

For an anthropologist's perspective, see for example, Forsythe and Buchanan (1988).

23. See Winograd (1987) on the way medical expert systems are becoming medical advisors. Lipscombe (1989) develops these points further.

Chapter 7

1. Coulter (1983) has an interesting discussion of Wittgenstein's idea of rules and its relationship to cognitive theory.

2. For a detailed example of the types of factors that can be considered see Collins (1985, chapter 3). There I describe what happened when my collaborator tried to

build a second TEA-laser, almost identical to one he had built some years earlier. When it did not work, we jointly began to explore what seem in retrospect the most bizarre possibilities for what might count as significant differences between the two lasers. See also this volume, chapter 9.

3. In the current state of the expert system art 100,000 is virtually as distant as the effectively infinite number that I suggest to comprise human competence. My guess would be that as machines become more powerful, and if the 100,000 rule machine becomes a possibility, then either human expertise will cease to be seen as a collection of rules, or the number of rules that are claimed as the human limit will increase dramatically so that it remains well over the horizon.

4. If it requires an indefinite, open-ended collection of rules to represent human culture, then complete success—whatever that means—is no more attainable than travel at the speed of light. Speedy improvement *may* not be possible either. If the number of new rules that are needed for each "unit increase" in cultural competence grows exponentially—and at a rate faster than improvements in computer power—then "small" advances will become harder and harder. It was Lighthill (1972) who invented the term "combinatorial explosion" to describe such a state of affairs.

Weizenbaum (1976) describes the way that computer hackers press onward, adding rule after rule to their programs, in the false belief that enough exceptions and special cases will eventually reproduce human behavior.

5. The song "Some Enchanted Evening" suggests that there is something special about a crowded room. The song goes on to suggest that even under these circumstances one will somehow know that one has met one's partner. The "somehow" catches the point nicely.

6. There are other ways to go wrong as a knowledge base grows. As the system reaches the limits of its abilities, there is no reason to suppose that the first mistake it makes will be a small one. It may get everything right to a point and then make catastrophic mistakes when it crosses the limits of its ability. To caricature, a medical expert system might be quite good at understanding the causes of backaches until it is confronted with someone with an embedded axe. The machine might well not be programmed to ask, "Is there anything stuck in your back?"—in which case its suggested treatment would be catastrophically inappropriate. Knowledge engineers aim for what they call "graceful degradation." *Disgraceful* degradation might be a better thing, for then it would be obvious to everyone when the machine had reached the limit of its abilities.

The opposite side of this problem is the computer that explains too much because it cannot match its performance to its user. Such a machine would be very tedious to use.

7. Festinger and his collaborators (1956) infiltrated a millennial cult secretively. They wanted to witness the cult's progress without influencing it by their actions. At one point, a collaborator was asked by a cult member to describe what he saw when contemplating. Determined not to reinforce the group in any way, the researcher replied that he saw nothing at all. But the cult member still invested the remark with great significance. "That is not nothing," she stated, "That is *the void*!!"

8. It *goes without saying*, as does almost everything else, that I have to know that beating eggs is not like beating rugs or children.

9. Such systems have disadvantages and advantages when compared to a book; it is not easy to read the rules from beginning to end, whereas on the other hand, the system finds its way swiftly to the right place in its knowledge base for the particular consultation in hand. One might say that the system acts as an intelligent index to its data base.

10. It should not be thought that Class I systems are always cheap and easy to build, nor guaranteed to succeed. A system built by British Gas to prescribe application of herbicides at remote pumping stations cost tens of thousands of pounds to build and, at the time of writing, is not in regular use. See Dreyfus and Dreyfus (1986) for a scathing critique of the claims of existing expert systems.

11. For a full account of the MYCIN story, see Buchanan and Shortliffe (1984).

12. These figures are taken by a talk given by Wiig at the expert system conference "ES85" in December 1985.

13. As I will go on to point out, another solution is for lay persons to become more accomplished in the esoterica of the domain, so that line b——b moves up.

14. Actually, even in such a simple case there is cause for doubt. In Britain, the convention is that a return ticket on a train is far less expensive than two singles. On a recent journey I wanted to travel from Bath to Heathrow and back. I wanted to travel there via London, Paddington Station, so that I could call in at the American Embassy in the morning and then return some days later more directly to Bath from Heathrow via Reading (see figure 7n1). The part of the journey from Paddington to Heathrow is done on an underground train, but the separate fares are consolidated in the charge made by British Rail. The journey from Heathrow to Reading is done by bus, but again, the whole fare from Heathrow to Bath is collected by British Rail to be distributed later.

The cost of the journey is different via the two routes, and the normal British Rail solution would have been to charge the mean of the two return fares. But the London route involves the independent carrier, London Transport, for the Paddington-Heathrow leg whereas the Reading route involves the independent bus company, National Express, for the Heathrow-Reading leg. British Rail and these independent carriers have no way of sharing out the cost of a return ticket where the two legs of the journey are conducted along the two different routes.

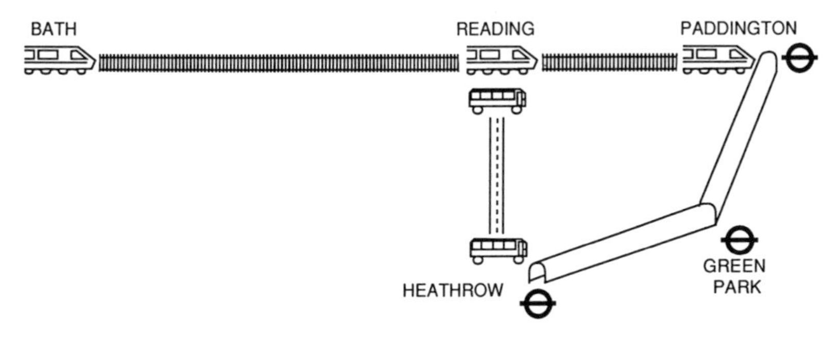

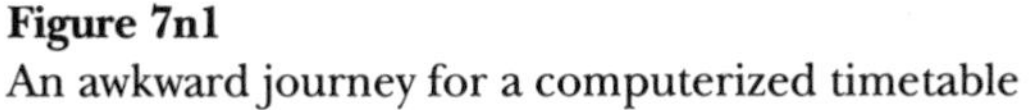
Figure 7n1
An awkward journey for a computerized timetable

The official solution was to buy two single tickets at exorbitant cost. We managed to work out, however, that I could spend less by throwing away part of my journey entitlement. I bought a return fare via Paddington and a single bus fare from Heathrow to Reading. On my way back I could use the bus ticket to get to Reading and board the train from London at Reading using the return half of the Paddington ticket without going to Paddington at all. This way I had to buy entitlements to travel that I would discard. These covered Heathrow to Paddington Station and Paddington to Reading. Throwing away part of my journey entitlement turned out to be much the least expensive solution. This is not the sort of solution one would expect to find in any straightforward railway information system.

What is worse, there was an added complication. The Paddington-Heathrow underground ticket can only be used from Paddington underground station but I needed to stop off at Green Park to go to the American Embassy in Grosvenor Square. The only way I could legitimately do this was to buy a separate underground ticket for travel from Paddington to Green Park and back to Paddington so that I could board at Paddington for the journey to Heathrow. Green Park, however, is well on the way to Paddington so this would be a waste of time and money. It would be much more sensible to buy a single underground ticket from Paddington to Green Park and board the train for the journey to Heathrow from there when I had completed my business at the American Embassy. The crucial question to British Rail was: "What chance have I of being stopped if I try to use my *Paddington* to Heathrow ticket from *Green Park*?" This is unambiguously against the rules. We concluded, using our understanding of human nature, that the risk was worthwhile and that sometimes it is necessary to cheat. As it turns out, I did complete the journey successfully.

No general rule about cheating could be programmed into an expert system. This was a particular journey under local circumstances. The rule cannot be generalized. For example, flying Air Canada with a ticket from Toronto to London I was refused admission to the flight in Montreal. I had to travel back to Toronto by train, taking about ten hours, board the same plane there, only to land once more and take off in Montreal an hour later. Anticipating such decisions requires us to understand national propensities for awkwardness, to have fine moral judgment, and to be able to estimate the likely outcome of individual personal interactions. Nevertheless, they *are* problems of timetabling.

15. Gilbert claims that irrespective of whether it can succeed in replacing DHSS counterclerks:

> [An] important reason for developing such systems is that they contribute, if only in a small way, to the distribution of information within our society which otherwise would be undemocratically locked into the hands of "experts". (Gilbert 1985, p. 2)

Whether this ambition is realized must depend on how the systems are implemented and used rather than the immanent properties of their design. (I am grateful to Nigel Gilbert for helpful discussion of the issues involved.)

Chapter 8

1. Tom Gieryn initially gave me the idea for looking at the instructions on Coca-Cola machines. He said he'd got it from Ruth Schwartz-Cowan, but she denies being the originator.

2. In the 1980s machine the locus of the esoteric has moved on to the details of scoring and reward. A series of instructions, begins thus:

Tilt penalty. Disqualifies ball in play from additional scoring.
Making S-H-A-R-P lanes, S-H-O-O-T-E-R Target, or Kickout Hole advance 2x- 3x- 4x- 5x-bonus.

I suppose there will come a time when these instructions will be redundant within the pinball community. Perhaps they already are.

3. Photographs of vintage "fruit machines" were found in Parker (n.d.).

4. I am grateful to Karl Scheele for taking time to guide me through the Smithsonian exhibits.

5. Non-Americans first encountering the Washington Metro will have discovered the esoteric nature of the ticket machines—soon mastered, but initially extraordinarily obscure. Suchman (1987) shows how difficult to interpret are the instructions on advanced Xerox machines. Even such simple machines still require human assistance in the face of the uncultured. My hypothesis is that such a complex system as the Washington Metro could only have been invented because Americans already knew the "language"of the ticket machines. Again, the Smithsonian may hold the secret. There a MAILOMAT postal machine of 1939 uses the same code:

First Direction: Deposit Coins for one or more letter. No change returned. [A display reveals the amount deposited.]
Second direction: Turn knob to stamp value needed for each letter. Maximum, 32 cents. [When you turn the knob, the value of the postage stamp appears in a little window.]
Third direction: Push letter in and let go, address side up, as shown below. [With a small reproduction of a letter to show which way up it should go as with the dollar bill on the Metro.]

There may be a thesis here on the vernacular of machines.

6. Modern studies of science show that in spite of this appearance science is really a local craft activity in many of its aspects (see chapter 1, note 4).

7. Shapin and Schaffer (1987, p. 29) write of Robert Boyle's troubles as he tried to build a vacuum pump with a glass receiver: "Given the state of the glass blower's art (which Boyle continually lamented), receivers were likely to crack and even to implode."

8. Another particularly interesting kind of upward transformation takes place when machinery and instruments are used to transform a manual and perceptual skill into something much more akin to a fact. For example, measurement of the temperature in a steel manufacturing process may change from a skilled judgment of color to a matter of reading a dial. This kind of process is usually described as involving the deskilling of workers. It is not really deskilling, however, for two different reasons. I have argued that the use of every piece of knowledge in the top box rests on the foundation of knowledge contained in the lower boxes and it is no different in this case. It is simply that the ability to tell the temperature of molten steel from its color is a narrowly distributed skill, whereas the ability to read a dial is a widely distributed skill—at least in Western societies. We do not value widely distributed skills, and we do not really notice that they are skills at all.

The second point to note about this type of movement is that it can only happen successfully if that part of the action that is replaced by machine, or whatever, is best performed as a machine-like act. Many attempts to replace humans with robots or, less ambitiously, with automated tools, founder on the belated discovery that what appeared to be a purely mechanical job required the abilities of humans to act in the full meaning of the term, if the task is to be completed efficiently. For noteworthy examples of the analysis of the amount of skill exercised in routine jobs, see Kusterer (1978), Jones (1984), Jones and Wood (1985), and Lazonick and Brush (1985). See also McKenna (1962) for an interesting semifictional account. In later chapters the same will be seen to be true of the chart in figure 8.2. It turns out that it could not be used in the mechanical way that I describe here. If it really is the case that there is an aspect of the task of the human operative that can be replaced by a mechanical device without loss, then what is being replaced is a machine-like action, and in this respect of the task, the human *must already have been deskilled.* Machines don't cause deskilling, only the organization of work can cause it.

9. Think of what has happened to our idea of the skill involved in, say, typewriting, or driving a car. These have become more and more negligible accomplishments. The change has a little to do with changing technology (cars are easier to drive in one sense, but much harder to drive on our crowded roads) but much to do with the wider distribution of these skills. The high status of the scribe in less-developed countries, where the skill of writing is still narrowly distributed, illustrates the point.

10. For the notion of "war stories" as the location of cultural skills, see Orr (1987).

11. As the paint-spraying robot is the embodiment of the space-time coordinates of paint spraying.

12. The same applies to the development of the 8 inch rule. Laser builders who knew about top leads only in terms of the 8 inch rule would not have the flexibility of expertise to, say, lengthen the lead appropriately if other parameters were changed. They would not be in a good position to experiment with new designs.

13. I will only discuss top lead length, but this can stand for all the knowledge involved in laser building.

Chapter 9

1. See Antaki (1988) for an extended discussion of the range of explanation.

2. Though not all rule trees would be suitable even for this purpose. For example, there are explanations that would be too complex for a human user to comprehend. Michie and Johnston (1985) point out that even if the workings of many computer programs could be rendered transparent to the end user, they would be incomprehensible. For example, chess-playing programs include large look-ahead trees, but a detailed account of the nodes and branches on such a tree would not help a chess player to understand what the computer was doing or to override its conclusions because there would be far too much information to use, and because the information would not correspond to human ways of thinking about chess. Michie and Johnston talk about a "human window" in computer programming—a window of reasoning that is like human reasoning in depth and complexity. Actually, computer reasoning can at best reflect only a special subset of human reasoning.

3. University of Surrey, March 20–21, 1986.

4. See, for example, Bobrow, Mittal, and Stefik (1986).

5. For a detailed discussion of perceptions of similarity and difference when applied to TEA-lasers see Collins (1985, chapter 4).

6. For references to other studies see chapter 1, note 3. For a more sophisticated attempt to explain scientific induction which nevertheless falls foul of the same objection, see Holland et al. (1986).

7. See Collins and Shapin (1988) for a discussion of the parallels between classroom science and real science. See Gorman (1987) for an interesting book review that makes many similar points.

8. This material was produced by Judy Palmer, a teacher of mathematics, as part of her assignment for the "sociology of scientific knowledge" unit on the Master of Education course at The University of Bath. She was asked to reconsider one of her typical mathematics lessons in the light of recent work in the philosophy and sociology of science and point out ways in which its pedagogical force depended on alternative interpretations of the material being hidden or ignored. I am grateful to Ms. Palmer for allowing me to use the material.

The other columns produced by the class follow; each row represents the work of one student:

a	b	c	a^2	b^2
7.1	9.0	11.3	50.4	81.0
5.0	7.0	8.4	25.0	49.0
8.5	7.9	11.4	72.3	62.4
3.7	5.1	6.1	13.7	26.0
8.4	7.5	11.4	70.6	56.0
6.0	8.0	9.9	36.0	64.0
9.4	8.3	12.6	88.4	68.9
10.1	9.5	13.8	102.0	90.3
7.6	8.3	11.1	57.8	68.9
8.2	9.2	12.3	67.2	84.6
11.3	10.7	15.7	127.7	114.5
10.4	9.6	14.3	108.1	92.2
9.3	10.4	13.8	86.5	108.2
5.2	7.8	9.7	27.0	60.8

9. For a more extended discussion of the BACON program see Collins (1989). The whole of that issue of *Social Studies of Science* is devoted to a symposium on BACON and its implications for the sociology of scientific knowledge.

10. Two, four, six, eight,
Who do we appreciate?
etc.

11. Of course my simple example is heavily loaded to illustrate my point because there are so few terms in the initial series. But given more terms there is no reason to suppose the machine induction will coincide with human induction.

12. Another method of automated knowledge gathering rests on the idea that much useful knowledge is contained in textbooks:

> Since the expert builds the knowledge base partly from the past experience and textbook cases, there is reason to hope that an induction program could build a knowledge base for an expert system in a similar way. (Buchanan et al. 1983, pp. 130–131)

> The bottleneck concerns getting the known volume of information into a computer representation, not getting yet more information out of experts. For example, if we could have textbooks automatically translated into computer representations maybe 80% of the perceived bottleneck would be resolved. (quoted in Forsythe 1987, p. 4)

Automated textbook browsing may well reduce the time spent by knowledge engineers on knowledge elicitation but even if it could be made to work there is no reason to suppose it would solve any other problems of expert systems. Indeed the knowledge base found in textbooks is likely to be more impoverished than that obtainable in conversation because textbooks tend to be stripped of heuristic knowledge.

Chapter 10

1. We concluded that crystal growing would make a suitable expert system project for a number of reasons: crystal growing does not require real-time control, and therefore there are no problems of processing speed nor difficulties to do with the system having to pick up from where it left off after a period of dormancy while slow events happen in the world; the process was fairly cheap and could be repeated many times; the cycle of repetition was only a few days; the expertise was on hand and an expert was ready and enthusiastic; the end result of the process was fairly clear (or so it seemed at the outset), either one has grown a crystal or one has not; finally, Green had once been an industrial chemist so the area would not be unfamiliar to him.

2. The work of crystal growing in the Bath department was begun by Brian Pamplin, editor of the book mentioned in the text. By coincidence this is the same Dr. Pamplin who had originated the "spoon-bending" experiments described in Collins and Pinch (1982), but he had nothing to do with the project described here. I am sorry to say that Dr. Pamplin died in 1987.

3. It is worth noting that the ampule method appears to provide the crystal grower with an easily controlled "closed universe." This, it will turn out, is far from the case.

4. In fact, an expert system to aid large-scale industrial crystal growth using the Czochralski method has been developed. See Wainwright (1987).

5. Green, it must be remembered, had not seen any crystal growing in practice, so he was not in a good position to guide Draper toward talking of his experience. I tried to keep as quiet as possible in the early parts of the knowledge elicitation, only interfering later as I saw that the system would never be built if the discussion continued in the same way.

6. As I have argued in earlier chapters, even extended talk represents only an infinitesimal part of what would have to be spoken if all of what one person knew was to be transferred to a completely ignorant person.

Chapter 11

1. Although I talk loosely of apprenticeship, my training, in fact, was comparatively vestigial. I underwent none of the hardships or intensity of relationship with the work that characterizes apprenticeship proper. For a description of a more full-blown apprenticeship see, for example, Cooper 1980.

2. At the time of writing this book I decided to try to teach myself to play the piano starting from a position of almost complete musical ignorance and never having had any significant contact with people who know anything of the performance of music. The same sort of useful surprise awaited me there. I started by trying to learn to play the piano in the same way as I learned to type. That is, I tried to teach my

fingers to respond to the notes written on the stave without intervention of the senses. I found that I could learn to do this by continual practice on simple pieces, but each new piece presented difficulties nearly as great as the first. The fingers had to be trained almost from the start for each new piece. It is as though I was learning to type by mastering first one sentence, then another, in the hope of eventually knowing all the sentences in the language. Fortunately, a musically socialized colleague in casual conversation told me that only the very finest pianists can play a piece without practice. Learners start by building up a repertoire of pieces rather than by training their fingers to respond to the notes. This makes learning to play the piano seem far more achievable.

3. The temporal sequence of this chapter is a little strange. I have pointed out that CRYSTAL's advice differed from Draper's method, but at the time I did not know what CRYSTAL would have advised; CRYSTAL was not yet built.

4. To see what could be deduced from reading the literature alone I asked Green to come up with a method for growing this crystal. After a few days of study—especially of Pamplin's book—he explained his reasoning to me as follows:

Green: My technique for dealing with this was to look up bismuth and arsenic in the periodic table at the back to see if they would give me any clues and I noticed that bismuth and arsenic are in the same group of the periodic table with nitrogen and phosphorous and this kind of thing, but not oxygen . . . And I looked them up in the inorganic compounds index and I found there was a bismuth oxide compound with silicon referred to on page 279 and arsenic was referred to in a different place. But I also came across a reference to arsenic in gallium arsenide in another article in the book. So, I thought, I'm on to something here. I came to the conclusion . . . that bismuth has a low melting point. And I found gallium arsenide here, and I came to the conclusion that—OK, there is sufficient clues for me to think that crystal pulling was an appropriate technique because it also says that "because this technique is a relatively fast crystal growth process it also finds wide application in the laboratory for synthesis in single crystal form of many new materials." So, I thought, well—and this is the kind of knowledge I shouldn't have used—I thought, it's highly unlikely that Harry and Bob will have indulged in something that is going to take three weeks to see the results so I reckon he used a technique that was relatively fast . . . I thought there were sufficient clues here to suggest that he did this by some kind of crystal-pulling technique.

Now I've not got enough here to work out what kind of furnace, etc. . . .

I then explained the problem of the condensing arsenic vapor making the vessel opaque.

Green: So you can't control the process visually

Collins: Yeah!

Green: [Laughs]

Collins: So that's something you wouldn't have thought of, is it, from reading that book.

Green: Absolutely.

5. Of course, a bigger project could have dealt with more of these cases. Full-scale expert systems are built over many person-years rather than a few person-weeks. I will have to rely on the reader's sympathy to extrapolate the difficulties we found in building our small system to full-scale work. I will have to rely on the reader's resisting the temptation to extrapolate *from* the solvability of the problems that I can describe in detail because of my *retrospective* understanding *to* the potential solubility of all such problems.

6. On this, my first observation of crystal growing, I merely watched as Draper weighed the materials—I learned to weigh later. I describe learning to weigh now as I want to indicate the details of the skills involved as we go along rather than back-tracking clumsily just to be faithful to the historical narrative.

Chapter 12

1. All the following was said in conversation between Green, Draper, and myself during the knowledge elicitation sessions.

Green: Let's start with boats.

Draper: Boats—you can probably roughly divide them into about four categories: glass, metal, ceramic and carbon, and sort of vitreous materials like boron nitride and vitreous carbon. So if you start off with the glassy ones, at the very bottom end, though I've never known anyone do it, you could start off with soda glass, which is a very low melting point glass. The next one up would be Pyrex glass, which would soften at about 500 at a rough guess. . . . In principle, if you had a very low melting point there is no reason why you shouldn't use an organic—waxy sort of material. . . . You could even use a plastic—you could use PTFE [Teflon].

Green: Would it be used here?

Draper: No. But I can think of some reasons you could use it. It would be a very good release agent. . . . If you did actually grow something in a PTFE container, it would release very easily.

Collins: But you've never actually used a PTFE container?

Draper: No, but I can think of some good reasons—if you wanted to grow a very low melting point compound and you wanted to make sure it would pop out easily, there is no reason why you shouldn't use something like PTFE. . . .

Draper: There is another glass, Vycor, which melts about 1,000, but we don't use it—we go straight to silica. . . . After that you're starting to get into the ceramics of one sort or another. Then it's usually a matter of price and availability. Alumina would be the most common and that would go up to, say, about 1,900.

Collins: Have you used that?

Draper: Not me personally, but some of my predecessors have. . . . There's a whole range. I think the other most common one is vitreous carbon and also vitreous boron nitride. Obviously you can only use the carbon in a vacuum, or a neutral or reducing atmosphere, because coal burns. So it's no good trying to grow oxides of crystals with an over-pressure of oxygen. . . . Vitreous carbon, say, maximum of 3,000.

Green: What about boron nitride?

Draper: 1,700 . . .

Collins: Have you used boron nitride?

Draper: No—I've seen catalogs of it. . . .

Green: What else, Bob?

Draper: There are a whole range of refractories. Things like thoria—thorium oxide, magnesia—magnesium oxide, zirconia, beryllia, alumina—we've had that—but in most cases you'd probably use alumina first of all.

Green: These all suffer from the problem of nucleation?

Draper: Yes.

Green: Is there any way round that?

Draper: It's a matter of making the inside smooth.

This discussion illustrates, as in the previous chapter, the problem of the way an expert's formal knowledge gets in the way of description of experience. Experience becomes invisible and inconsequential, compared to the formal technical knowledge that is so highly valued.

2. For a detailed discussion of the way artifacts are separated from genuine phenomena in microscope-based research, see Lynch (1985).

3. New and exotic etching agents are always being invented in the laboratory. For example, while we were considering our procedure a graduate student announced that he had invented a solution of 10% bromine in ethyl alcohol. He was clearly quite proud of it.

4. For the difference between a demonstration and a real system, see Alvey, Meyers, and Greaves (1987).

5. Clancey (1987) provides an up-to-date account of an experiment to use an expert system as a tutor. Clancey's experiment shows that the mere exercise of writing out a rule base demonstrates unspoken assumptions within it (such as the order in which diagnostic hypotheses are tested). One other thing that the research discovered was the need to include justifications for rules to provide, as Clancey (1987, p. 223) states, "an understanding that allows the problem solver to violate the rules in unusual situations!"

6. It was very interesting that most of our users were somewhat embarrassed when CRYSTAL asked them for the probable vapor pressures of the intended reaction. They felt that they *ought* to know the answer. Few of them were ready to say that the vapor pressure was unknowable, and most were apologetic and took a guess. Their respect for the textbook version of knowledge and for the computer program, which seemed to know what it was talking about—after all, it was teaching them things they did not know—was disturbing.

Chapter 13

1. For a treatment of Turing's personal interest in the matter of confusion between the sexes, see Hodges (1985).

The idea of the imitation game helped one of my colleagues to see the point of Searle's Chinese room argument. If a man can completely imitate a woman's conversational competence, does this mean he has the mind of a woman?

2. It is worth thinking about the way this subtle feature of the Turing Test relates to intelligence tests in general. These, of course, are notorious for an apparent universality that masks cultural specificity. My argument is exactly the opposite. I take human abilities—that which the designers of intelligent machines strive to mimic—to be essentially a matter of the ability to participate in a social group. The Turing Test in its initial formulation hides this capacity and reinforces an intelligence test-like definition of intelligence.

3. For a similar argument to that of Neumaier regarding the inadequacy of the Turing Test, see Sokolowski (1988).

4. See chapter 15, note 2, for an interchange with a partly mechanized telephonist.

5. For example, the Israeli showman Uri Geller was said by critics of his performance to have a radio embedded in his tooth. Eliminate this, and other possibilities like it, and there is always the possibility of coconspiracy with the experimenters to fall back on.

6. For this argument worked out in detail for the case of gravitational radiation, see Collins (1985, chapter 4). For the implications in the case of testing new technologies, see Constant (1983), and MacKenzie (1988).

7. The cynic, or sophisticate, will notice that I press the Turing Test as an adequate test of intelligence because of its interactive element. I *want* intelligence to be seen as something having to do with social interaction. That is the basis of my whole argument. Further, I adjust the protocol so as to make socialization a salient feature.

8. Another feature of this test is that the judges looked at ready-made transcripts,

they did not interact with the computer or patient in real time. So-called Turing Tests on medical expert systems often compare computer diagnoses and human diagnoses based on medical transcripts rather than examination of a real patient. This loads the dice heavily in favor of the computer. The most well-known such test was that done on the medical expert system MYCIN. For a recent repetition of the same protocol, see "Italian Medical Advisor Takes the Turing Test," *Intelligence*, Texas Instruments No. 15, April, 1989.

9. Colby's judges were, presumably, trying not to be fooled. Some people, however, are actually more respectful of machines than they would be of humans because of what they have heard of the power of computers. Also there are circumstances when people prefer to "converse" with an impersonal machine than with a human being about delicate matters:

> Nearly 100 [police] forces in the US and Canada [have] successfully elicited information from children unwilling to confide in police or social workers [about sexual abuse].... "Even the best human interrogators may send involuntary negative messages through body language, voice inflection or facial expressions," said James Norton, a robot operator with the Alaskan police. "The advantage of Ares is that it has no obvious sex, is ethnically versatile and cannot possibly resemble a perpetrator of child abuse".... Its success is attributed to its ability to create an atmosphere of intimate friendship, according to Tom Zaken of 21st Century Robotics." (*Sunday Times*, September 6, 1987)

This preference for robots in some social interactions might mean that computers will gain in acceptability as surrogate human beings; it could mean that they are more likely to be treated as human-like. (For a discussion of the way children relate to computers, see Turkle [1984].)

10. This sort of process is already well advanced. In some areas a "scientific problem" is becoming defined as something that can be handled by a computer. For example, the mainframe computer has had a significant impact on the social sciences, devaluing microscopic interpretative methods and overvaluing large-scale quantitative surveys.

11. For an approach that encourages the reassessment of ourselves as adjuncts to the world of technology, see Latour (forthcoming). Latour treats nonanimate things as belonging to our sociocognitive networks.

12. For "distance lends enchantment," see chapters 11 and 12 of this book and Collins (1985), especially pp. 144–145.

13. We should not, however, neglect the frightening ability of humans to work in the other direction. Creatures who do function identically to us have been redefined as less than human—as objects unfit to share our forms-of-life. I refer to the ideology of racial persecution.

14. Thus, a human whose responses were "wooden," as though they had learned only the behavioral coordinates of action not the meaning of the acts themselves, would make a poor control. We do come across people who seem to have something of this approach to life. People who go through the niceties in a bureaucratic mechanical way as though they had learned how to behave from a book on etiquette or by observation from a distant planet. According to Searle, the biochemistry of the brain is a necessary condition of consciousness and understanding. It does not, however, seem to be a sufficient condition. Those who fail to see the point of the Searle argument may have in mind the class of people who do seem to struggle through life on the basis of a largely calculative approach.

15. It is hard to remain clear about the probabilities involved in the Turing Test.

First, think of an ordinary human matched against an ultimately bad program such as, say, a talking clock. Here the interrogator's probability of correct identification is 100%. As the machine gets better the probability falls. One might think it could fall no lower than 50% because this is what could be achieved by guesswork if no clues could be read from the teletype responses. But imagine programs matched against people with very "wooden" responses. Many interrogators will mistake a "wooden" performer for a computer so the probability of making the correct identification falls toward zero, not because the machine is like a human but because the human is too like a machine.

16. Last time I made a telephone call from San Francisco airport a voice told me that the number I had called was unobtainable and then went on to tell me that I was speaking to a recorded message. The recording was very good and the *justified* assumption was that I would not otherwise know I was being spoken to by a machine. Of course, if my suspicions were aroused it would not be hard to discover the nonhuman identity of this particular switchboard operator. The point is that a Turing Test is not interesting from our point of view if the interrogator does not know it is a test.

17. In the imitation game the modern-day problem arises out of the similarity in the socialization of modern men and women. In our small experiment at Bath we found that the sort of answers that seemed to give a clue to sex were to do with gender stereotyping. Thus, a man pretending to be a woman who was asked about which film stars he found attractive tended to produce stereotypes of masculinity. Real women gave much less obvious answers, the authenticity of which could be spotted by the women but not by the men.

18. As we investigate whether or not dolphins really are intelligent we try to *understand* what they say— to make a social contact between their world and ours. To try to determine the intelligence of culturally distant entities *in their terms* is as futile as trying to design a culturally neutral intelligence test. The same point shows the futility of the idea that interacting machines might live within their own world of intelligence impenetrable to us. The same might be true of wastepaper baskets. This is not to say that it is not morally appropriate to assume the intelligence of living things that do not share our culture even if it cannot be proved.

Chapter 14

1. In discussing Block's ideas in this way I am doing violence to the author's intentions. Block wanted to use the idea to distinguish between intelligence and a "conduit" for intelligence in which all the intelligence exhibited by the machine was possessed by the programmer. For this purpose the machine need only perform once at a satisfactory level. It seems Block had something similar to Searle's argument in mind in writing his paper. I argued that Searle's hypothesis directed attention away from the interesting questions of artificial intelligence. In the same way, I am not interested in Block's argument but only in his design. I believe his design cannot do all that he claims for it. His imaginary program cannot mimic the performance of a human as thoroughly as he thinks. Showing why this is so is interestingly and revealingly difficult.

2. I adopt the name "Block machine" for my refined version in deference to Block's original design of "string searcher."

3. In Collins (1985, p. 23) I used a joke relying on onomatopoeia as an example of an impossible task for a speech transcriber. The Block machine would have no trouble with the joke so long as the programmer understood it.

4. A machine of the *power* of the Block machine could be a perfect "chess" player if programmed deterministically—there are *only* about 10^{120} possible moves in chess!

5. Though, as we will see, it would not be proof against linguistic tricks.

6. This is what makes confidence tricksters so interesting. They have to use a range of techniques to get people to accept that they are what they are not. The crucial point is that the *mark*, or the trickster's victim, must be charitable—must want to believe what the trickster is saying. The mark does most of the work.

7. We do not want the machine to mimic someone who is mimicking someone else. This breaks a rule of protocol discussed in the last chapter.

8. In Block's original article he simply has programmers allowing responses that "Aunt Bertha" might make, without thinking through the ramifications of this. If only a single set of possible responses is catered for, without allowing any conversational variation, then the machine can produce only quasi-human behavior.

9. Either within a run or between separate runs.

10. Whenever I have presented these ideas, members of the audience have invariably come up with clever modifications that seem to overcome the problems I outline. For example, each string could be labeled with the *probability* of its use within the personality being modeled. Thus while there would be as many strings expressing conservative sentiments as expressing progressive sentiments, where the programmer was a progressive thinker each conservative string would have a very low probability of selection attached to it so that the chance of selection of a conservative string summed across all the conservative strings would be very low. The problem is that this requires that we know the relevant probabilities. That is, if I am the programmer, I have to be able to articulate, in the form of a set of statistics, the characteristics of my potential conversational performance. But if I can do that, then the major problems of artificial intelligence are already solved. Innovatory rule-breaking behavior will have turned out to be just a matter of statistics.

Another way around the problem is simply to equip the machine with a set of defensive responses. Thus, the machine simply need not make political statements at all—all strings that include a political sentiment on the part of the machine are excluded, whereas political expressions by the interrogator are met with responses such as, "I never discuss politics"; strings that contain nonsense expressed by interrogators (see below), are met with defensive responses such as, "I don't really follow that." Undoubtedly such a strategy would make for an acceptable conversational partner but not one that could mimic everyone's competence. For example, it would not be able to mimic a trained Turing-Test interrogator or an exciting conversationalist.

All the technical fixes of this type that I have encountered require either that we make explicit our tacit knowledge of how to carry on conversations, returning us to the problem that the Block design so cleverly side-steps, or they fail to reproduce conversational competence.

11. Actually, the general form of this piece of nonsense is taken from the BBC Radio series of the 1950s, "The Goon Show." The piece is not entirely free of form, therefore. For example, 60 *x*'s would not represent a turn, and, in fact, 1B has only certain permissible rhythms.

12. As before, we could "hack" around this problem by dividing the memorized strings into sensible ones and nonsense-containing ones and introducing a rule about the frequency with which the nonsense-containing ones can be sampled. This, however, is open to all the previous objections against hacking round the political preference problem.

13. Because innovative social behavior may involve being grammatically inaccurate from time to time it is not even clear any longer that we can capture grammatical capacity in the machine without being unduly restrictive.

14. In passing we might ask where Lynch and the other conversational analysts get their conventions from. How does Lynch know how to spell phonetically and what warrant does he have for thinking that this spelling better represents the sound of his respondents?

15. The same argument applies to onomatopoeia. Though I have argued that the machine could cope with a joke that depended on onomatopoeia, the fact that onomatopoeic representation of sounds is less restricted than standard English spelling means that the joke and its minor variants would be repeated rather more frequently among the conversation strings than corresponds to its frequency of repetition in ordinary English. This means the machine would have rather more of a tendency to tell onomatopoeic jokes than a normal person!

16. This is what leads Neumaier and other critics to undervalue the test and consider "extended Turing Tests," involving robots and so forth. I will argue that there is no need for such elaboration. The performance of a skill can be examined even via teletypes.

17. I am now predicting what will happen when I spell check this chapter. . . . I am now confirming that the prediction was correct. Unsurprisingly, the spell checker had a lot of trouble with all the nonstandard words in this chapter.

18. A dictionary of acceptable pairings of words would be much bigger than a regular dictionary and so forth. The Block machine is much less puzzling when it is thought of as a giant "sense checker" and its mistakes are compared to those of a regular spell checker.

Chapter 15

1. See Lipscombe (1989) for an illuminating discussion of the attempt to replace skill with hyper-accuracy in the development of medical expert systems.

2. I find myself continually surprised by ingenuity of this sort. Once I needed to find the telephone number of the Day's Inn in Phoenix, Arizona. I was told by the operator that the Day's Inn listing now referred callers to the San Carlos hotel. It seemed its name had been changed. I asked for the number of the San Carlos, but the operator told me there were many hotels of that name in Phoenix; she asked for the address, which I did not know. Then she discovered that one San Carlos was listed as the "Day's Inn San Carlos." "That must be the one," we agreed. That kind of awkward inquiry could not have been automated. At that point, however, the

operator left me for another inquirer while a mechanical voice spoke the number. The creative search was done by the human but a little time was saved by using a machine to do that part of the task that could be executed with a machine-like action.

3. Comparing figure 15.1 with figure 8.1, the conventional hierarchy of knowledge types, facts, rules, and heuristics are all to be found in box 2 of figure 15.1. The contents of boxes, 1, 3, and 4 of figure 15.1 are all found among the manual and perceptual skills of figure 8.1. The contents of box 4 may move up the hierarchy of knowledge types in the appropriate circumstances.

4. Sherry Turkle, in a paper presented to the Bath Science Studies Centre, suggests that some of the excitement over neural nets arises because of the romantic connotations of their hidden and impenetrable mechanism. In this respect they appear as mysterious as the brain.

5. Box 1 is problematic. Because one cannot fully express the rules pertaining to a form-of-life, it might be thought that this box should be empty. The distinction between boxes 1 and 3 is simply that in the former case one may be paying some attention to what one is doing even if that attention does not stretch to every aspect of the act. For example, as new ways-of-going-on in science are being established, great attention is paid to every implication. Once a new scientific object has been established, however, the associated culture becomes our taken-for-granted-reality.

6. The theory of machine-like action adds nothing new in revealing how these mistakes are made. What it does add is an understanding of the nature of the areas of action where the decoding does work. It shows, then, why so many are tempted to extend the method in inappropriate ways and why they are so frustrated with blanket philosophical criticism. It is a successful method in some areas of human action. The critics do not notice this.

7. I speak of Draper making the necessary contributions, but it must be remembered that Draper was in continual touch with the crystal-growing community. Draper had always to bear in mind what would count as a satisfactory crystal in the circumstances. In fact, the crystal was being grown for experimental use in a laboratory at MIT. Draper took advice on the matter whenever he was uncertain himself. He tested his instincts out on the community.

8. Building NEOCRYSTAL would be an attempt to shift our expert system from Class II to Class III.

9. I once failed in my attempt to change the oil filter on my car—one of the simplest jobs there is. An oil filter looks roughly like a can of beans. It is screwed into the engine and must be rotated in a counterclockwise direction to remove it. My filter would not turn so I used a trick I had read about in a handyman's magazine. I hammered a screwdriver through the soft body of the oil filter so as to use it as a lever. This, the magazine assured me, would enable me to unscrew the filter easily. Unfortunately, all that happened was that the metal of the filter tore. I now had an immobile car and a potentially very large bill. A friend eventually showed me how to use a special tool and a great deal of strength to finish the job.

References

Alvey, P., Myers, C. D., and Greaves, M. F. (1987) "High Performance Expert Systems: I. Escaping From the Demonstrator Class." *Medical Information* 12 (2): 85–95.

Anderson J. R. (1982) "Acquisition of a Cognitive Skill." *Psychological Review* 89 (4): 369–406.

Antaki, C. ed. (1988) *Analysing Lay Explanation: A Casebook of Methods.* London: Sage.

Atkinson, R., and Delamont, S. (1977) "Mock-ups and Cock-ups: The Stage Management of Guided Discovery Instruction." In *School Experience: Explorations in the Sociology of Education*, edited by P. Woods and M. Hammersley. London: Croom Helm.

Baker, G. P., and Hacker, P. M. S. (1985) *An Analytical Commentary on the Philosophical Investigations.* Oxford: Blackwell.

Barnes, B., and Edge, D. (eds.) (1982) *Science in Context: Readings in the Sociology of Science*. Milton Keynes: Open University Press.

Berry, D. (1987) "The Problem of Implicit Knowledge." *Expert Systems: The International Journal of Knowledge Engineering* 4 (3): 144–151.

Block, N. (1981) "Psychologism and Behaviourism." *The Philosophical Review* 90: 5–43.

Bloomfield, B. (1986) "Capturing Expertise by Rule Induction." *Knowledge Engineering Review* 1 (4): 30–36.

Bloor, D. (1973) "Wittgenstein and Mannheim on the Sociology of Mathematics." *Studies in the History and Philosophy of Science* 4: 173–191.

Bloor, D. (1976) *Knowledge and Social Imagery.* London: Routledge and Kegan Paul.

Bloor, D. (1983) *Wittgenstein: A Social Theory of Knowledge.* London: Macmillan.

Bloor, D. (1987) "The Living Foundations of Mathematics." *Social Studies of Science* 17 (2): 337–357.

Bobrow, D. G., Mittal, S., and Stefik, M. J. (1986) "Expert Systems: Perils and Promise." *Communications of the ACM* 2 (9): 880–894.

Boden, M. (1985) Paper presented to Expert Systems 85 conference, University of Warwick, December.

Buchanan, B., Barstow, D., Bechtal R., Bennett, W., Clancey, W., Kulikowski, C., Mitchell, T., and Waterman, D. (1983) "Constructing an Expert System."In *Building Expert Systems*, edited by F. Hayes-Roth, D. Waterman, and D. Lenat. Reading, Mass.: Addison-Wesley.

Buchanan, B. , and Shortliffe, E. (1984) *Rule-Based Expert Systems.* Reading, Mass.: Addison-Wesley.

Clancey, W. J. (1984) "Methodology for Building an Intelligent Tutoring System." In *Methods and Tactics in Cognitive Science*, edited by W. Kintsch. Hillsdale, N.J.: Lawrence Erlbaum.

Clancey, W. J. (1987) *Knowledge-Based Tutoring: The GUIDON Program.* Cambridge, Mass.: MIT Press.

Colby, K. M. (1981) "Modeling a Paranoid Mind," *The Behavioural and Brain Sciences* 4: 515–560.

Colby, K. M., Hilf, F. D., Weber, S., and Kraemer, H. C (1972) "Turing-Like Indistinguishability Tests for the Validation of a Computer Simulation of a Paranoid Process." *Artificial Intelligence* 3: 199–221.

Collins, H. M. (1974) "The TEA Set: Tacit Knowledge and Scientific Networks." *Science Studies* 4: 165–86. (Reprinted in Barnes and Edge 1982).

Collins, H. M. (1975) "The Seven Sexes: A Study in the Sociology of a Phenomenon, or the Replication of Experiments in Physics." *Sociology* 9: 205–224. (reprinted in Barnes and Edge, 1982).

Collins, H. M. (ed.) (1981a) *Knowledge and Controversy: Studies of Modern Natural Science*, Special Issue *of Social Studies of Science* 11 (1).

Collins, H. M. (1981b) "Son of the Seven Sexes: The Social Destruction of Physical Phenomenon." In Collins 1981a: pp. 33–62.

Collins, H. M. (1982) *Sociology of Scientific Knowledge: A Sourcebook.* Bath: Bath University Press.

Collins, H. M. (1985) *Changing Order: Replication and Induction in Scientific Practice.* London and Beverly Hills: Sage.

Collins, H. M. (1989) "Computers and the Sociology of Scientific Knowledge." *Social Studies of Science* 19: 613–624.

Collins, H. M., Green, R., and Draper, R. (1985) "Where's the Expertise: Expert Systems as a Medium of Knowledge Transfer." In *Expert Systems 85*, edited by M. J. Merry. Cambridge: Cambridge University Press, pp. 323–334.

Collins, H. M., and Harrison, R. (1975) "Building a TEA Laser: The Caprices of Communication." *Social Studies of Science* 5: 441–445.

Collins, H. M., and Pinch, T. J. (1982) *Frames of Meaning: The Social Construction of Extraordinary Science.* London: Routledge and Kegan Paul.

Collins, H. M., and Shapin, S. (1988) "Experiment, Science Teaching and the New History and Sociology of Science." In *Test-Tube History: Teaching the History of Science and Technology*, edited by M. Shortland and A. Warwick. (First published in Proceedings of the International Conference on Using History of Physics in Innovatory Physics Education, Pavia, Italy, September, 5–9, 1983.).

Constant, E. W. (1983) "Scientific Theory and Technological Testability: Science, Dynamometers, and Water Turbines in the 19th Century." *Technology and Culture* 24: 183–198.

Cooley, M. (1987) *Architect or Bee: The Human Price of Technology.* London: The Hogarth Press.

Cooper, E. (1980) *The Wood Carvers of Hong-Kong: Craft Production in the World Capitalist Periphery* . Cambridge: Cambridge University Press.

Coulter, J. (1983) *Rethinking Cognitive Theory.* London: Macmillan.

Crossfield, L. P. (1986) "Explanation in Regard to a Welfare Benefits Advice System." Paper presented to Explanation in Expert Systems conference, University of Surrey, March 20–21.

Crothers, D. (1987) "How to pot black with some angular momentum and a touch of gravity." *The Guardian,* August 27, 1987.

Davis, R. (1982) "Expert Systems: Where Are We? and Where Do We Go From Here?" MIT Artificial Intelligence Laboratory Memo No. 665, June 1982, Cambridge, Mass.

Dodson, D. C. (1986) "Explanation Explanation. . . A Brief Sketch." Paper presented to Explanation in Expert Systems conference, University of Surrey, March 20–2.

Dreyfus, H. (1979) *What Computers Can't Do.* New York: Harper and Row.

Dreyfus H. L, and Dreyfus, S. E. (1986) Mind Over Machine: *The Power of Human Intuition and Expertise in the Era of the Computer.* New York: Free Press.

Dreyfus, H. L., and Dreyfus, S. E. (1988) "Making a Mind Versus Modeling the Brain: Artificial Intelligence at a Branch Point." *Daedalus* 117: 15–44.

Earman, J., and Glymour, C. (1980) "Relativity and Eclipses: The British Eclipse Expeditions of 1919 and their Predecessors." *Historical Studies in the Physical Sciences* 11 (1): 49–85.

Evans, J. St. B. T., and Wason, P. C. (1976) "Rationalisation in a Reasoning Task." *British Journal of Psychology* 67: 479–486.

Feigenbaum, E., and McCorduck, P. (1984) *The Fifth Generation.* London and Sydney: Pan Books.

Festinger, L., Riecken, H. W., and Schachter, S. (1956) *When Prophecy Fails.* New York: Harper.

Feyerabend, P. K. (1975) *Against Method.* London: New Left Books.

Fitts, P. M. (1964) "Perceptual-Motor Skill Learning." In *Categories of Human Learning,* edited by A. W. Melton. New York: Academic Press.

Fleck, L. (1979) *Genesis and Development of a Scientific Fact.* Chicago: University of Chicago Press (first published in German in 1935).

Forsythe, D. (1987) "Engineering Knowledge: An Anthropological Study of an Artificial Intelligence Laboratory." Paper presented at the 12th Annual Meeting of the Society for Social Studies of Science, Worcester, Mass., November 19–22.

Forsythe, D. E., and Buchanan, B. G. (1988) "An Empirical Study of Knowledge Elicitation: Some Pitfalls and Suggestions." In *Methods in Knowledge Engineering,* special issue of *IEEE Transactions on Systems, Man and Cybernetics* , edited by P.E. Lehner and L. Adelman.

Fulda, J. S. (1988) "An Expert System for an Idiosyncratic Domain: Love, Intimacy and Friendship." *SIGART Newsletter* No. 105, July, pp. 30–34.

Garfinkel, H. (1967) *Studies in Ethnomethodology.* Englewood Cliffs, N.J.: Prentice-Hall.

Gellner, E. (1974) "The New Idealism: Cause and Meaning in the Social Sciences." In *Positivism and Sociology* , edited by A. Giddens. London: Heinemann.

Gilbert, N. (1985) "Computer Help With Welfare Benefits." *Computer Bulletin* 1 (4): 2–4.

Gilbreth, F. B., and Gilbreth, L. M. (1919) *Applied Motion Study.* New York: Macmillan.

Gorman, M. E. (1987) "Will the Next Kepler Be a Computer?" *Science and Technology Studies* 5 (2): 63–65.

Green, R., and Wood, S. (1963) "Expert Knowledge Elicitation—Or How to Grow Trees." In *Expert Systems 83*, Proceedings of the British Computer Society Specialist Group on Expert Systems Annual Conference.

Hamilton, W. (trans.) (1973) Plato, *The Phaedrus.* Harmondsworth: Penguin.

Harper, D. (1987) *Working Knowledge: Skill and Community in a Small Shop.* Chicago: University of Chicago Press.

Harris J. R. (1976) "Skills, Coal and British Industry in the Eighteenth Century." *History* 61: 167–182.

Harvey, B. (1981) "Plausibility and the Evaluation of Knowledge: A Case Study in Experimental Quantum Mechanics." In Collins 1981a: pp. 95–130.

Haugeland, J. (1985) *Artificial Intelligence: The Very Idea.* Cambridge, Mass.: MIT Press.

Hayes-Roth, F. (1985) "Knowledge-Based-Systems—The State of the Art in The US." In *The Knowledge Engineering Review* 1: 18–27.

Heiser, J. F., Colby, K. M., Faught, W. S., and Parkinson, R. C. (1979) "Can Psychiatrists Distinguish A Computer Simulation of Paranoia From the Real Thing?: The Limitations of Turing-Like Tests as Measure of the Adequacy of Simulations." *Journal of Psychiatric Research* 15:149–162.

Hodges, A. (1985) *Alan Turing: The Enigma of Intelligence* . London: Unwin.

Holland, J. H., Holyoak, K. J., Nisbett, R. E., and Thagard, P. R. (1986), *Induction: Processes of Inference, Learning and Discovery.* Cambridge, Mass.: MIT Press.

Holton, G. (1978) *The Scientific Imagination.* Cambridge: Cambridge University Press.

Hughes, S. (1986) "HOW and WHY: HOW Far Will They Take Us, and WHY Should We Need Any More?" Paper presented to Explanation in Expert Systems conference, University of Surrey, March 20–21.

Jackson, P. (1986) "Explaining Expert System Behaviour." Paper presented to Explanation in Expert Systems conference, University of Surrey, March 20–21.

Jackson, P., and Lefevre, P. (1984) "On the Application of Rule-Based Techniques to the Design of Advice Giving Systems." *International Journal of Man-Machine Studies.* 20: 63–86.

Jeremy, D. (1981) *Transatlantic Industrial Revolution.* Cambridge, Mass.: MIT Press.

Jones, B. (1984) "The Division of Labour and the Distribution of Tacit Knowledge in the Automation of Metal Machining." In *Design of Work in Automated Manufacturing Systems,* edited by T. Martin. Oxford: Pergamon Press.

Jones, B. (1989) "When Certainty Fails: Inside the Automatic Factory." In *The Transformation of Work?,* edited by S. Wood. London: Unwin-Hyman.

Jones, B., and Wood, S. (1985) "Tacit Skills, Division of Labour and New Technology." *Sociologie du Travail:* 407–422.

Kidd. A., and Welbank, M. (1984) "Knowledge Acquisition." In *Expert Systems 'State of the Art' Report,* edited by J. Fox. Oxford: Pergamon Infotech Ltd.

Knapp, B. (1963) *Skill in Sport: The Attainment of Proficiency.* London: Routledge and Kegan Paul.

Knight, J. A. (1986) "Explanation in a Co-Operative Expert Assistant." Paper presented to Explanation in Expert Systems conference, University of Surrey, March 20–21.

Knorr-Cetina, K. D. (1981) *The Manufacture of Knowledge.* Oxford: Pergamon Press.

Kuhn, T. S. (1961). "The Function of Measurement in Modern Physical Science." *ISIS* 52: 162–176.

Kuhn, T. S. (1962) *The Structure of Scientific Revolutions.* Chicago: University of Chicago Press.

Kusterer, K. C. (1978) *Know-How on the Job: The Important Working Knowledge of "Unskilled" Workers.* Boulder: Westview Press.

Lakatos, I. (1976) *Proofs and Refutations.* Cambridge: Cambridge University Press.

Langley, P., Simon, H. A., Bradshaw, G. L., and Zytkow, J. M. (1987) *Scientific Discovery: Computational Explorations of the Creative Process.* Cambridge, Mass.: MIT Press.

Latour, B. (1987), *Science in Action.* Milton Keynes: Open University Press.

Latour, B. (forthcoming) "Sociology of a Door." Paper prepared for Twente II conference, Enschede, Holland, September 1987.

Latour, B., and Woolgar, S. (1979) *Laboratory Life: The Social Construction of Scientific Facts.* London and Beverly Hills: Sage.

Lave, J. (1986) "The Values of Quantification." In *Power, Action and Belief: A New Sociology of Knowledge,* edited by J. Law. Sociological Review Monograph No. 32. London: Routledge.

Lave, J. (1988) *Cognition in Practice.* Cambridge: Cambridge University Press.

Lazonick, W., and Brush, T. (1985) "The 'Horndal Effect' in Early U. S. Manufacturing." *Explorations in Economic History* 22:53–96.

Leith, P. (1986) "Fundamental Errors in Logic Programming." *The Computer Journal 29* (6):545–554.

Lenat, D., Prakash, M., Shepherd, M. (1986) "CYC: Using Common Sense Knowledge to Overcome Brittleness and Knowledge Acquisition Bottlenecks." *The AI Magazine* 6(4): 65–85.

Lighthill, J. (1972) "Artificial Intelligence: A General Survey." In *Artificial Intelligence: A Paper Symposium.* London: Science Research Council.

Lipscombe, B. (1989) "Expert Systems and Computer-Controlled Decision Making in Medicine." *AI and Society* 3:184–197.

Lynch, M. (1985) *Art and Artifact in Laboratory Science: A Study of Shop Work and Shop Talk in a Research Laboratory.* London: Routledge and Kegan Paul.

McClelland, J. L., Rumelhart, D. E., and the PDP Research Group (1986) *Parallel Distributed Processing: Explorations in the Microstructure of Cognition.* Cambridge, Mass.: MIT Press.

McCorduck, P. (1979) *Machines Who Think.* San Francisco: W. H. Freeman.

McDermott, D. (1981) "Artificial Intelligence Meets Natural Stupidity." In *Mind Design,* edited by J. Haugeland. Cambridge, Mass.: MIT Press, pp. 143–160.

McKenna, R. (1962) *The Sand Pebbles.* New York: Harper & Row.

McKenzie, D. (1988) "From Kwajelein to Armageddon?: Testing and the Social Construction of Missile Accuracy." In *The Uses of Experiment,* edited by D. Gooding, T. Pinch, and S. Schaffer. Cambridge: Cambridge University Press.

McKeown, K. R. (1980) "Generating Relevant Explanations: Natural Language Responses to Questions About Data-Base Structure." *Proceedings of the First Annual National Conference on Artificial Intelligence,* AAAI, pp. 306–309.

Michie, D., and Johnston, R. (1985) *The Creative Computer: Machine Intelligence and Human Knowledge.* Harmondsworth: Pelican.

Neumaier, O. (1987) "A Wittgensteinian View of Artificial Intelligence." In *Artificial Intelligence: The Case Against,* edited by Rainer Born. London: Croom Helm, pp. 132–173.

Oldman, D., and Drucker C. (1985) "The Non-Reduceability of Ethno- Methods: Can People and Computers Form a Society. In *Social Action and Artificial Intelligence,* edited by G. Gilbert and C. Heath. Aldershot: Gowe.

Olson, J. R., and Rueter, H. H. (1987) "Extracting Expertise from Experts: Methods for Knowledge Acquisition." *Expert Systems* 4(3):152–168.

Orr, J. (1987) "Talking About Machines: Social Aspects of Expertise," (mimeo).

Pamplin, B. R. (ed.) (1980) *Crystal Growth.* 2nd ed. Oxford: Pergamon Press.

Papert, S. (1988) "One AI or Many?" *Daedalus* 117:1–14.

Parker, S. (n.d.) *Slot Machines.* Las Vegas: Nevada Publications,p. 45.

Pickering, A. (1981) "Constraints on Controversy: The Case of the Magnetic Monopole." In Collins 1981a: pp. 63–93.

Pinch, T. J. (1981) "The Sun-Set: The Presentation of Certainty in Scientific Life." In Collins 1981a: pp. 131–158.

Pinch, T. J. (1986) *Confronting Nature: The Sociology of Solar-Neutrino Detection.* Dordrecht: Reidel.

Polanyi, M. (1958) *Personal Knowledge.* London: Routledge and Kegan Paul.

Pollner, M. (1974) "Mundane Reasoning." *Philosophy of the Social Sciences* (4):35–54.

Price, D. J. (1989) "The Advantages of Doubling in Dutch." *The Guardian,* April 20, 1989.

Reber, A. S., Kassin, S. M., Lewis, S., and Cantor, G. (1980) "On the Relationship Between Implicit and Explicit Modes of Learning a Complex Rule Structure." *Journal of Experimental Psychology: Human Learning and Memory* 6:492–502.

Rose, M. J. (1988) *Industrial Behaviour: Research and Control.* Harmondsworth: Penguin Books.

Rudwick, M. (1985) *The Great Devonian Controversy: The Shaping of Scientific Knowledge Among Gentlemanly Specialists.* Chicago: Chicago University Press.

Rymaszewski, R. (1986) "Learner Perspectives for Effective Explanation." Paper presented to Explanation in Expert Systems conference, University of Surrey, March 20–21.

Schutz, A. (1964) *Studies in Social Theory, Collected Papers, vol. II.* The Hague: Martinus Nijhoff.

Searle, J. (1980) "Minds, Brains and Programs." *The Behavioural and Brain Sciences* (3):417–424.

Searle, J. (1984) *Minds, Brains and Science.* Cambridge, Mass.: Harvard University Press.

Shah, J. S. (1980) "Zone Refining and Its Applications." In *Crystal Growth,* edited by B. R. Pamplin. Oxford: Pergamon Press, pp. 301–355.

Shapin, S., and Schaffer, S. (1985) *Leviathan and the Air Pump: Hobbes, Boyle and the Experimental Life.* Princeton: Princeton University Press.

Shortliffe, E. H., Davis, R., Axline, S. G., Buchanan, B. G., Green, C. C., and Cohen, S. N. (1975) "Computer Based Consultations in Clinical Therapeutics: Explanation and Rule Acquisition Capabilities of the MYCIN System." *Computers and Biomedical Research* 8:303–319.

Shurkin, J. N. (1983) "Expert Systems: The Practical Face of Artificial Intelligence." *Technology Review* 86(6):72–78.

Sokolowski, R. (1988) "Natural and Artificial Intelligence." *Daedalus: Proceedings of the American Academy of Arts and Sciences* 117(1):45–64.

Styles, J. (1980) "Our Traitorous Money Makers." In *An Ungovernable People,* edited by J. Styles and J. Brewer. London: Hutchinson.

Suchman, L. A. (1987) *Plans and Situated Action: The Problem of Human-Machine Interaction.* Cambridge: Cambridge University Press.

Sudnow D. (1978) *Ways of the Hand: The Organisation of Improvised Conduct.* London: Routledge and Kegan Paul.

Taylor, F. W. (1947) *Testimony Before the Special House Committee.* New York: Harper and Row.

Travis, G. D. L. (1981) "Replicating Replication? Aspects of the Social Construction of Learning in Planarian Worms." In Collins 1981a: pp. 11–32.

Turing, A. M. (1982) "Computing Machinery and Intelligence." Reprint. *The Mind's I,* edited by D. Hofstadter and D. Dennet. Harmondsworth: Penguin, pp. 53–66.

Turkle, S. (1984) *The Second Self: Computers and the Human Spirit*. New York: Simon and Schuster.

Wainwright, J. (1987) "Crystal Tips." *Expert Systems User* 3(7):24–28.

Waterman, D. A., and Newell, A. (1971) "Protocol Analysis as a Task for Artificial Intelligence." *Artificial Intelligence* 2:285318.

Webb, J. (1988) *The Guardian*, March 12, 1988.

Weizenbaum, J. (1974) Communication in *SIGART Newsletter*, July 1974.

Weizenbaum, J. (1976) *Computer Power and Human Reason: From Judgement to Calculation*. San Francisco: W. H. Freeman.

Welbank, M. (1983) "A Review of Knowledge Acquisition Techniques for Expert Systems." Martlesham Heath: British Telecom Research Laboratories.

Wilson, B. (ed.) (1970) *Rationality*. Oxford: Blackwell.

Winch, P. G. (1958) *The Idea of a Social Science*. London: Routledge and Kegan Paul.

Winograd, T. (1987) "Thinking Machines: Can There Be?, Are We?" CSLI Report No. 87/100. Stanford University.

Winograd, T., and Flores, F. (1986) *Understanding Computers and Cognition: A New Foundation for Design*. Norwood, N. J.: Ablex.

Wittgenstein, L. (1953) *Philosophical Investigations*. Oxford: Blackwell.

Wittgenstein, L. (1956) *Remarks on the Foundations of Mathematics*. Oxford: Blackwell.

Woolgar, S. (1985) "Why Not a Sociology of Machines?: The Case of Sociology and Artificial Intelligence." *Sociology* 19:557–572.

Wynne, B. (1988) "Unruly Technology: Practical Rules, Impractical Discourses and Public Understanding." *Social Studies of Science* 18:147–167.

Index